Measurements in Quantum Mechanics

Measurements in Quantum Mechanics

Editor

Fred Gajos

scitus
academics

Measurements in Quantum Mechanics
Edited by **Fred Gajos**

ISBN: 978-1-68117-318-4
Library of Congress Control Number: 2016937630

www.scitusacademics.com
Box No. 4766, 616 Corporate Way,
Suite 2, Valley Cottage,
NY 10989

Printed in United States of America on Acid Free Paper

Preface

A measurement always causes the system to jump into a specific state of the dynamic variable measured, the proper value of which is equal to the measurement result. Measuring plays an important role in quantum mechanics and is seen in various ways between different interpretations of quantum mechanics. Quantum mechanics is the very small physics branch. This leads to some very weird conclusions about the physical world. Many of the equations of classical mechanics, which describe how things move at everyday dimensions and speeds, are no longer useful on the scale of atoms and electrons. In classical mechanics, objects exist at a certain time in a certain place. Quantum mechanics is perhaps seen as the most remarkable development in the physics of the 20th century. Every successful theory deals solely with "measurement results." Quantum mechanics is completely different from classical measurement physics, because direct measurement as a classical form is impossible in the microscopic world of quantum mechanics. This book focuses on the difficult part of quantum mechanics in measurements.

Table of Contents

1

CHAPTER

A NEW FORMULATION OF QUANTUM MECHANICS

A. I. Arbab[1], Faisal A. Yassein[2]

[1]*Department of Physics, Faculty of Science, University of Khartoum, Khartoum, Sudan*

[2]*Department of Physics, Faculty of Science, Alneelain University, Khartoum, Sudan*

ABSTRACT

A new formulation of quantum mechanics based on differential commutator brackets is developed. We have found a wave equation representing the fermionic particle. In this formalism, the continuity equation mixes the Klein-Gordon and Schrodinger probability density while keeping the Klein-Gordon and Schrodinger current unaltered. We have found time and space transformations under which Dirac's equation is invariant. The invariance of Maxwell's equations under these transformations shows that the electric and magnetic fields of a moving charged particle are perpendicular to the velocity of the propagating particle. This formulation agrees with the quaternionic formulation recently developed by Arbab.

Keywords: Mathematical Formulation; Quantum Mechanics; Differential Commutator Brackets

1. INTRODUCTION

Schrodinger's equation was used to explain and describe all phenomena in atomic physics. However, after the development of the theory of special relativity by Einstein, there was a need to unify quantum mechanics and special relativity into a single Relativistic Quantum Theory. Despite the success of Schrodinger's equation in describing quite accurately the Hydrogen spectrum and giving correct predictions for a large amount of spectral data, this equation is not invariant under Lorenz transformations. In other words Schrodinger's equation is not relativistic and is only an approximation valid at the non-relativistic limit when the velocities of the particles involved are much smaller than the speed of light.

Quantum mechanics has been formulated by assigning an operator for any dynamical observable. In Heisenberg formalism, the operator is governed by a commutator bracket. The fundamental commutator bracket relates to the position, and momentum is given by $[x, p_x] = i\hbar$. The commutator bracket generalizes the Poisson bracket of classical mechanics. If an operator commutes with Hamiltonian of the system, then the dynamical variable corresponding to that operator is said to be conserved. An equation compatible with Lorentz transformation guarantees its applicability to any inertial frame. Such an equation is symmetric in space-time. Thus a symmetric space-time formulation of any theory will generally guarantee the universality of the theory. However, Schrodinger equations doesn't exhibit this feature because it is not symmetric in space and time. To remedy this problem, Klein and Gordon looked for an equation which is second order in space and time and consequently obtained the Klein-Gordon equation (KG). The probability density in this theory is found to be non-positive definite. Consequently, Dirac thought for a linear equation in space and time that has no such a problem. He obtained the familiar Dirac equation with a positive definite probability. However, the probability in KG formalism is later on (from a theoretic field point of view) interpreted as a charge density rather than a probability density which could be positive or negative [1-3].

With these motivation, we adopt a differential commutator bracket involving first order space and time derivative operators to formulate the Maxwell equations and quantum mechanics. This is in addition to our recent quarternionic formulation of physical laws, where we have shown that many physical equations are found to emerge from a unified form of physical variables [4]. Moreover, using quaternions, we have recently shown that quantum mechanics can be formulated in a set of three equations [5]. In such a formulation, the Dirac and Klein-Gordon equations emerge from a set of three equations obtained from the application of an eigen-value problem of the linear momentum.

We aim in this paper to derive the equation of motion of the quantum system by applying the vanishing differential commutator brackets. It is interesting to note that these commutator brackets are Lorentz invariant.

Moreover, $[\Delta t,\Delta x]=[\Delta t',\Delta x']$, where t is time, $\nabla=\frac{\partial}{\partial x}\hat{i}+\frac{\partial}{\partial y}\hat{j}+\frac{\partial}{\partial z}\hat{k}$, $\Delta x=x_2-x_1$ and t',x' are the moving time and space coordinates. We know that the second order partial derivatives commute for space-space variables. We don't assume here that this property is a priori for space and time. To guarantee this, we eliminate the time derivative of a quantity that is acted by a space (∇) derivative followed by a time derivative, and vice versa. In expanding the differential commutator brackets, we don't commute time and space derivative, but rather eliminate the time derivative by the space derivative, and vice versa. These linear differential commutator brackets may enlighten us to quantize these physical quantities. By employing the differential commutator brackets of the vector A and scalar potential ϕ, we have derived Maxwell equations without invoking any a priori physical law [6]. We would like here to apply the differential commutator brackets to explore quantum mechanics.

2. DIFFERENTIAL COMMUTATORS ALGEBRA

Define the three linear differential commutator brackets as follows [6];

$$\left[\frac{\partial}{\partial t},\nabla\right]=0,\quad \left[\frac{\partial}{\partial t},\nabla\cdot\right]=0,\quad \left[\frac{\partial}{\partial t},\nabla\times\right]=0, \tag{1}$$

where ∇ and $\frac{\partial}{\partial t}$ are the space and time derivatives.

For a scalar ϕ and a vector A, one finds that:

$$\left[\frac{\partial}{\partial t},\nabla\right]\phi=\frac{\partial}{\partial t}(\nabla\phi)-\nabla\left(\frac{\partial\phi}{\partial t}\right), \tag{2}$$

and

$$\left[\frac{\partial}{\partial t},\nabla\cdot\right]A=\frac{\partial}{\partial t}(\nabla\cdot A)-\nabla\cdot\left(\frac{\partial A}{\partial t}\right),$$
$$\left[\frac{\partial}{\partial t},\nabla\times\right]A=\frac{\partial}{\partial t}(\nabla\times A)-\nabla\times\left(\frac{\partial A}{\partial t}\right). \tag{3}$$

Moreover, one can show that:

$$\left[\frac{\partial}{\partial t},\nabla\cdot\right](\phi\boldsymbol{A})=\phi\left\{\left[\frac{\partial}{\partial t},\nabla\cdot\right]\boldsymbol{A}\right\}+\left\{\left[\frac{\partial}{\partial t},\nabla\right]\phi\right\}\cdot\boldsymbol{A},$$
$$\left[\frac{\partial}{\partial t},\nabla\times\right](\phi\boldsymbol{A})=\phi\left\{\left[\frac{\partial}{\partial t},\nabla\times\right]\boldsymbol{A}\right\}+\left\{\left[\frac{\partial}{\partial t},\nabla\right]\phi\right\}\times\boldsymbol{A},$$
$$\left[\frac{\partial}{\partial t},\nabla\cdot\right](\boldsymbol{A}\times\boldsymbol{B})=\boldsymbol{B}\cdot\left\{\left[\frac{\partial}{\partial t},\nabla\times\right]\boldsymbol{A}\right\}-\boldsymbol{A}\cdot\left\{\left[\frac{\partial}{\partial t},\nabla\times\right]\boldsymbol{B}\right\}. \quad (4)$$

The differential commutator bracket satisfies the distribution rule:

$$\left[\hat{A}\hat{B},\hat{C}\right]=\hat{A}\left[\hat{B},\hat{C}\right]+\left[\hat{A},\hat{C}\right]\hat{B}, \quad (5)$$

where $\hat{A},\hat{B},\hat{C}$ stand for either ∇ or $\frac{\partial}{\partial t}$.

It is evident that the differential commutator brackets identities follow the same ordinary vector identities. We call the three differential commutator brackets in Equation (1) the grad-commutator bracket, the dotcommutator bracket, and the cross-commutator bracket, respecttively. The prime idea here is to replace the time derivative of a quantity by the space derivative ∇ of another quantity, and vice versa, so that the time derivative of a quantity is followed by a time derivative with which it commutes. We assume here that space and time derivatives don't commute. With this minimal assumption, we have shown here that all physical laws are determined by vanishing differential commutator bracket.

3. THE CONTINUITY EQUATION

Using quaternionic algebra [7-9], we have recently found that generalized continuity equations can be written as [5].

$$\nabla\cdot\boldsymbol{J}+\frac{\partial\rho}{\partial t}=0, \quad (6)$$

$$\nabla\left(\rho c^2\right)+\frac{\partial\boldsymbol{J}}{\partial t}=0, \quad (7)$$

and

$$\nabla \times \boldsymbol{J} = 0, \tag{8}$$

where c, ρ and $\boldsymbol{J}$ are the speed of light, current density, and probability density, respectively. Now consider the dot-commutator bracket of $\rho\boldsymbol{J}$.

$$\left[\frac{\partial}{\partial t}, \nabla\cdot\right](\rho\boldsymbol{J}) = \frac{\partial}{\partial t}(\nabla\cdot(\rho\boldsymbol{J})) - \nabla\cdot\left(\frac{\partial(\rho\boldsymbol{J})}{\partial t}\right) = 0. \tag{9}$$

Using Equations (6-8), and the vector identities

$$\begin{aligned} &\nabla\cdot(\phi\boldsymbol{G}) = (\nabla\phi)\cdot\boldsymbol{G} + \phi(\nabla\cdot\boldsymbol{G}), \\ &\nabla\times(\nabla\times\boldsymbol{G}) = \nabla(\nabla\cdot\boldsymbol{G}) - \nabla^2\boldsymbol{G} \end{aligned} \tag{10}$$

one obtains

$$\begin{aligned} &\left[\frac{\partial}{\partial t}, \nabla\cdot\right](\rho\boldsymbol{J}) \\ &= c^2\rho\left(\frac{1}{c^2}\frac{\partial^2\rho}{\partial t^2} - \nabla^2\rho\right) + \left(\frac{1}{c^2}\frac{\partial \boldsymbol{J}}{\partial t^2} - \nabla^2\boldsymbol{J}\right)\cdot\boldsymbol{J} = 0. \end{aligned} \tag{11}$$

For arbitrary ρ and $\boldsymbol{J}$, Equation (11) yields the two wave equations

$$\frac{1}{c^2}\frac{\partial^2\rho}{\partial t^2} - \nabla^2\rho = 0, \tag{12}$$

and

$$\frac{1}{c^2}\frac{\partial \boldsymbol{J}}{\partial t^2} - \nabla^2\boldsymbol{J} = 0. \tag{13}$$

Equations (12) and (13) are also obtained utilizing the quaternionic formulation following reference [5]. Hence, the wave equations of ρ and $\boldsymbol{J}$, in our present brackets formulation, are equivalent to

$$\left[\frac{\partial}{\partial t}, \nabla\cdot\right](\rho\boldsymbol{J}) = 0. \tag{14}$$

Equations (12) and (13) show that the charge and current densities satisfy a wave traveling at the speed of light in vacuum. It is remarkable to know that these two equations are already obtained in reference [5].

4. QUANTUM MECHANICS

Consider a particle described by the four vector $\Psi = \left(\frac{i}{c}\psi_0, \boldsymbol{\psi}\right)$. This is equivalent to spinor representation of ordinary quantum mechanics. We have recently developed a quaternionic quantum mechanics dealing with such a four vector [7-9]. The evolution of this four vector is given by the three equations [7-9]

$$\nabla \cdot \boldsymbol{\psi} - \frac{1}{c^2}\frac{\partial \psi_0}{\partial t} - \frac{m_0}{\hbar}\psi_0 = 0, \tag{15}$$

$$\nabla \psi_0 - \frac{\partial \boldsymbol{\psi}}{\partial t} - \frac{m_0 c^2}{\hbar}\boldsymbol{\psi} = 0, \tag{16}$$

and

$$\nabla \times \boldsymbol{\psi} = 0, \tag{17}$$

where m_0 and $\hbar$ are the quasi-particle mass and Planck constant, respectively. Equations (15-17) yield the two wave equations [5].

$$\frac{1}{c^2}\frac{\partial^2 \boldsymbol{\psi}}{\partial t^2} - \nabla^2 \boldsymbol{\psi} + 2\left(\frac{m_0}{\hbar}\right)\frac{\partial \boldsymbol{\psi}}{\partial t} + \left(\frac{m_0 c}{\hbar}\right)^2 \boldsymbol{\psi} = 0, \tag{18}$$

and

$$\frac{1}{c^2}\frac{\partial^2 \psi_0}{\partial t^2} - \nabla^2 \psi_0 + 2\left(\frac{m_0}{\hbar}\right)\frac{\partial \psi_0}{\partial t} + \left(\frac{m_0 c}{\hbar}\right)^2 \psi_0 = 0. \tag{19}$$

Using the transformation,

$$\frac{\partial}{\partial \tau} = \frac{\partial}{\partial t} + \frac{m_0 c^2}{\hbar}, \tag{20}$$

so that Equations (15) and (16) become,

$$\nabla \cdot \psi = \frac{1}{c^2}\frac{\partial \psi_0}{\partial \tau}, \quad \nabla \psi_0 = \frac{\partial \psi}{\partial \tau}. \tag{21}$$

Employing Equation (20), Equations (18) and (19) are transformed into the wave equations

$$\Pi^2 \psi_0 = 0, \;\; \Pi^2 \psi = 0\,, \; \Pi^2 = \frac{1}{c^2}\frac{\partial^2}{\partial \tau^2} - \nabla^2. \tag{22}$$

Equations (18) and (19) can be obtained from the Einstein's energy equation by setting $E' = E + im_0 c^2$, where $E'^2 = p^2 c^2$ and using the familiar quantum mechanical operator replacements, viz., $\hat{p} = -i\hbar\nabla$ and $\hat{E} = i\hbar\frac{\partial}{\partial t}$. $E' = cp$ is an equation for a massless particle. This is also evident from Equation (22). Thus, it is interesting that a massive particle can be transformed into a massless particle using Equation (20). Since energy is a real quantity, this equation is physically acceptable if it describes a particle with imaginary mass. In this case the energy equations split into two parts; one with $E' = E + m_0 c^2$ and the other with energy $E' = E - m_0 c^2$. Such energies can describe the state of a particle and antiparticle. A hypotheticcal particle with an imaginary mass moving at a speed higher than the speed of light in vacuum is known as tachyon [10]. Hence, our above equation can be used to treat the motion of tachyons. This implies that our equations, Equations (24) and (27) can be applied to tachyons. Some scientists propose that neutrino can be a tachyonic fermion [11]. We know that the Cherenkov radiation is emitted from a particle moving in a medium with a speed larger than the speed of light in vacuum. When the speed exceeds the speed of light in a vacuum, the extra energy acquired by the particle is transformed in radiation. This can happen momentarily for a particle keeping its total energy conserved. Thus, the excess energy (speed) is such that it compensates the dissipations.

Now consider the cross-commutator bracket of $\psi_0 \psi$,

$$\left[\frac{\partial}{\partial t}, \nabla \times\right](\psi_0 \psi) = \frac{\partial}{\partial t}\left(\nabla \times (\psi_0 \psi)\right) - \nabla \times \left(\frac{\partial(\psi_0 \psi)}{\partial t}\right) = 0. \tag{23}$$

Using Equations (15), (16), and (17), and the vector identities,

$$\nabla\times(\phi\boldsymbol{G})=(\nabla\phi)\times\boldsymbol{G}+\phi(\nabla\times\boldsymbol{G}), \tag{24}$$

yield the wave equation,

$$\frac{1}{c^2}\frac{\partial^2\boldsymbol{\psi}}{\partial t^2}-\nabla^2\boldsymbol{\psi}+2\left(\frac{m_0}{\hbar}\right)\frac{\partial\boldsymbol{\psi}}{\partial t}+\left(\frac{m_0c}{\hbar}\right)^2\boldsymbol{\psi}=0. \tag{25}$$

Similarly, the dot-commutator bracket of $\psi_0\boldsymbol{\psi}$,

$$\left[\frac{\partial}{\partial t},\nabla\cdot\right](\psi_0\boldsymbol{\psi})=\frac{\partial}{\partial t}(\nabla\cdot(\psi_0\boldsymbol{\psi}))-\nabla\cdot\left(\frac{\partial(\psi_0\boldsymbol{\psi})}{\partial t}\right)=0. \tag{26}$$

Upon using Equations (10), (15), and (16), one obtains the wave equation of ψ_0,

$$\frac{1}{c^2}\frac{\partial^2\psi_0}{\partial t^2}-\nabla^2\psi_0+2\left(\frac{m_0}{\hbar}\right)\frac{\partial\psi_0}{\partial t}+\left(\frac{m_0c}{\hbar}\right)^2\psi_0=0. \tag{27}$$

It is interesting to see that Equations (25), and (27) are the same as Equations (18), and (19) obtained from quaternionic manipulation. We thus write Equations (25), and (27) as,

$$\left[\frac{\partial}{\partial t},\nabla\times\right](\psi_0\boldsymbol{\psi})=0,\quad \left[\frac{\partial}{\partial t},\nabla\cdot\right](\psi_0\boldsymbol{\psi})=0. \tag{28}$$

5. DIRAC'S EQUATION

Dirac's equation can be written in the form [1-3],

$$\frac{1}{c}\frac{\partial\psi}{\partial t}+\boldsymbol{\alpha}\cdot\nabla\psi+\frac{im_0c\beta}{\hbar}\psi=0. \tag{29}$$

Consider the differential commutator bracket

$$\left[\frac{\partial}{\partial t},\boldsymbol{\alpha}\cdot\nabla\right]\psi=\frac{\partial}{\partial t}(\boldsymbol{\alpha}\cdot\nabla\psi)-(\boldsymbol{\alpha}\cdot\nabla)\frac{\partial\psi}{\partial t}=0. \tag{30}$$

Using Equation (29), Equation (30) yields,

$$\frac{1}{c^2}\frac{\partial^2\psi}{\partial t^2}-\nabla^2\psi+2\left(\frac{m_0 i\beta}{\hbar}\right)\frac{\partial\psi}{\partial t}-\left(\frac{m_0 c}{\hbar}\right)^2\psi=0, \tag{31}$$

where we have used the fact that $\beta=\begin{pmatrix}1 & 0\\ 0 & -1\end{pmatrix}$, $\alpha^2=\beta^2=1$ $\alpha=\begin{pmatrix}0 & \sigma\\ \sigma & 0\end{pmatrix}$, and σ are the Pauli matrices. Equation (31) can be obtained from Equation (29) by squaring it. This equation can be compared with the KleinGordon equation of spin-0 particles

$$\frac{1}{c^2}\frac{\partial^2\psi}{\partial t^2}-\nabla^2\psi+\left(\frac{m_0 c}{\hbar}\right)^2\psi=0.$$

Equation (31) is another form of Dirac's equation exhibiting the wave nature of spin-1/2 particles explicitly. Using the transformation,

$$\frac{\partial}{\partial\eta}=\frac{\partial}{\partial t}+i\frac{m_0 c^2}{\hbar}\beta, \tag{32}$$

Equation (31) can be written as,

$$\frac{1}{c^2}\frac{\partial^2\psi}{\partial\eta^2}-\nabla^2\psi=0. \tag{33}$$

This is a wave equation for a massless particle. Thus, a particle annihilates (loses its mass) after a time interval of $\Delta t=\frac{\hbar}{m_0 c^2}$ and then created (acquired a mass). It is interesting to notice that during such a period of time, energy can be violated as endorsed by the Heisenberg's uncertainty relation ($\Delta t\Delta E\geq\hbar$)where $\Delta E=2m_0c^2$. This also applies to the particles as defined by Equation (20). Equation (31) describes the behavior of a particle of a definite mass m_0. After a time of $\Delta t=\frac{\hbar}{m_0 c^2}$ the particle becomes a wave with energy $E'=pc$ governed by Equation (33). The particle interacts with the vacuum in such a way that when the particle becomes a wave (annihilates) gives its mass energy to the vacuum, and restores it after a time of Δt as defined before becoming a particle once again. This is the essence of the oscillatory motion as known as zitterbewegung motion [1-3]. This result supports the fact that there is a vacuum fluctuation associated with the particle. This means when a particle becomes a wave it gives its mass to the vacuum and restores it when becomes a corpuscule. Thus, the corpuscular and wave nature (duality) of a particle is concomitant with the particle motion. Since ψ is a four components spinor, we can write it in terms of two components doublets, viz., $\psi = \begin{pmatrix} \psi_+ \\ \psi_- \end{pmatrix}$. Substituting these decomposed spinors in Equation (31), one obtains the two equations,

$$\frac{1}{c^2}\frac{\partial^2 \psi_+}{\partial t^2} - \nabla^2 \psi_+ + 2\left(\frac{m_0 i}{\hbar}\right)\frac{\partial \psi_+}{\partial t} - \left(\frac{m_0 c}{\hbar}\right)^2 \psi_+ = 0, \tag{34}$$

and

$$\frac{1}{c^2}\frac{\partial^2 \psi_-}{\partial t^2} - \nabla^2 \psi_- - 2\left(\frac{m_0 i}{\hbar}\right)\frac{\partial \psi_-}{\partial t} - \left(\frac{m_0 c}{\hbar}\right)^2 \psi_- = 0, \tag{35}$$

Equations (34) and (35) imply two energy solutions, one with $E' = E - m_0 c^2$ and the other with energy $E' = E + m_0 c^2$. This is also evident from using the Einstein energy-momentum equation ($E^2 = p^2 c^2 + m_0^2 c^4$). The two energy states may define a particle and an antiparticle. Since the time factor in the wavefunction is of the form $\exp(-iEt/\hbar)$, the new wavefunction with the new time (η) will become $\exp(-iE'\eta/\hbar)$, where

$$\eta = t - i\frac{m_0 c^2}{\hbar}, \tag{36}$$

is a complex time, as evident from Equation (33). It can be seen as a rotation of the real time by a phase into a complex plane. Such an effect arises from the very nature of the particle when propagating in space-time. The third term in Equation (31) represents a dissipation that may result from the motion of the particle in space (ether). Hence, any massive particle should exhibit this sort of propagation when travels in space-time. This term is vanishingly small compared with the mass term in Equation (31) but very fundamental. Moreover, Our Equations (25) and (27) are equivalent to Dirac equations, Equations (34) and (35), if we replace m_0 by $-im_0$.

Consider now the case when ψ is space independent so that Equation (31) becomes,

$$\frac{\mathrm{d}^2\psi}{\mathrm{d}t^2} + 2\left(\frac{m_0 c^2 i\beta}{\hbar}\right)\frac{\mathrm{d}\psi}{\mathrm{d}t} - \left(\frac{m_0 c^2}{\hbar}\right)^2 \psi = 0. \tag{37}$$

with $\psi = \begin{pmatrix} \psi_+ \\ \psi_- \end{pmatrix}$ this yields the two equations

$$\frac{d^2\psi_+}{dt^2} + 2\left(\frac{m_0 c^2 i}{\hbar}\right)\frac{d\psi_+}{dt} - \left(\frac{m_0 c^2}{\hbar}\right)^2 \psi_+ = 0, \tag{38}$$

and

$$\frac{d^2\psi_-}{dt^2} - 2\left(\frac{m_0 c^2 i}{\hbar}\right)\frac{d\psi_-}{dt} - \left(\frac{m_0 c^2}{\hbar}\right)^2 \psi_- = 0. \tag{39}$$

These two equations have an oscillatory behavior, i.e.,

$$\psi_+(t) = A_+ \exp(-i\omega t), \quad \psi_-(t) = A_- \exp(i\omega t), \tag{40}$$

where $\omega = \frac{m_0 c^2}{\hbar}$ and A_+, A_- are constants. This means that the particle with the wavefunction ψ has two energy eigen states, one for a particle and the other one for an antiparticle. Hence, Equation (40) reveals that the particle is described by a standing wave having positive and negative energy. This is the essence of Dirac's theory. The two states are separated by an amount of energy, $\Delta E' = 2m_0 c^2$.

Using Equation (29), Equation (31) can be written in the form,

$$\frac{1}{c^2}\frac{\partial^2 \psi}{\partial t^2} - \nabla^2\psi - 2\left(\frac{m_0 c i}{\hbar}\right)\beta\boldsymbol{\alpha}\cdot\nabla\psi + \left(\frac{m_0 c}{\hbar}\right)^2 \psi = 0, \tag{41}$$

this can be written as

$$\frac{1}{c^2}\frac{\partial^2 \psi}{\partial t^2} - \nabla'^2\psi = 0, \tag{42}$$

where

$$\nabla' = \nabla + i\frac{m_0 c\beta}{\hbar}\boldsymbol{\alpha}. \tag{43}$$

Equation (42) is a wave equation in the new coordinate defined by Equation (43). Equation (43) can be written as

$$\boldsymbol{p}' = \boldsymbol{p} + m_0 c\beta\boldsymbol{\alpha}. \tag{44}$$

This can be compared with the covariant derivative that results from the interaction of a particle with a photon field $\boldsymbol{A}$, viz., $\boldsymbol{p}' = \boldsymbol{p} - e\boldsymbol{A}$. Equation (32) can be written as

$$E' = E - m_0 c^2 \beta . \tag{45}$$

Equations (32) and (43) can be combined into a single equation as

$$D_\mu = \partial_\mu + i\frac{m_0 c}{\hbar}\gamma_\mu, \quad \gamma_\mu = (\beta, \beta\alpha). \tag{46}$$

We call here the derivative D'_μ the spinor derivative. With this derivative the Dirac equation takes the simple forms,

$$P\psi = 0, \quad \text{where} \quad P_\mu = i\hbar\gamma^\mu D_\mu, \tag{47}$$

It is interesting to notice that Equations (47) looks like massless Dirac equation. The second term in Equation (44) represents a self interaction of the particle due to its spin. Since the vector potential is a gauge field, the spin of the particle will accordingly become a gauge quantity. In present case, the electron interacts with its spin that is related to $\boldsymbol{\alpha}$. This effect represents a self interaction of the particle. The algebra of the P_μ's commutator bracket is

$$\left[P^\mu, P^\nu\right] = 2m_0 c i\sigma^{\mu\nu}, \quad \sigma^{\mu\nu} = \frac{i}{2}\left[\gamma^\mu, \gamma^\nu\right]. \tag{48}$$

Thus, unlike partial derivative, spinor derivatives do not commute. The momenta commute for a massless particle. Equation (41) describes a particle with definite mass which after a characteristic distance of $\ell = \frac{\hbar}{m_0 c}$ becomes a wave as described by Equation (42). Hence, the corpuscular nature of the particle is exhibited after a distance of $X_0 = \frac{2\hbar}{m_0 c}$, and the wave nature after a time of $T_0 = \frac{2\hbar}{m_0 c^2}$. The particle's velocity must be in such a way to reach the next point in the same time required to be in the other state. This requires its velocity to be $\boldsymbol{v} = \boldsymbol{\alpha} c$. Thus, the particle remains in a continuous dual state (particle + wave). This duality is manifested during a time of T_0 at a distance of X_0. This may usher into a quantization of space and time in units of T_0 and X_0 as fundamental units. With the definition $\boldsymbol{v} = \boldsymbol{\alpha} c$, Dirac's equation can be written as

$$\frac{\partial}{\partial t} + v \cdot \nabla = -\frac{i m_0 c^2}{\hbar} \beta = \frac{\mathrm{d}}{\mathrm{d}t} \tag{49}$$

which implies that

$$i\hbar \frac{\mathrm{d}\psi}{\mathrm{d}t} = m_0 c^2 \beta \psi. \tag{50}$$

Hence, Equation (50) is a variant form of Dirac's equation. But since ψ can be written as two-components column, viz., $\psi = \begin{pmatrix} \psi_+ \\ \psi_- \end{pmatrix}$, the above equation implies that

$$i\hbar \frac{\mathrm{d}\psi_+}{\mathrm{d}t} = m_0 c^2 \psi_+, \quad i\hbar \frac{\mathrm{d}\psi_-}{\mathrm{d}t} = -m_0 c^2 \psi_-. \tag{51}$$

This shows that the operator $\hat{A} = i\hbar \frac{\mathrm{d}}{\mathrm{d}t}$ is the rest mass energy operator of the individual spinor components.

The Continuity Equation

Taking the complex conjugate of Equation (31) and multiplying it by ψ once from right, and subtract it from Equation (31) after multiplying it by ψ^* from left, we obtain the continuity equation,

$$\frac{\partial \rho_T}{\partial t} + \nabla \cdot \boldsymbol{J} = 0, \tag{52}$$

where

$$\rho_T = \rho_S + \rho_{KG}, \rho_S = \beta\psi^*\psi,$$
$$\rho_{KG} = \frac{i\hbar}{2m_0c^2}\left(\psi^*\frac{\partial\psi}{\partial t} - \psi\frac{\partial\psi^*}{\partial t}\right),$$
$$\boldsymbol{J} = \frac{\hbar}{2m_0 i}\left(\psi^*\nabla\psi - \psi\nabla\psi^*\right). \tag{53}$$

It is understood here that ψ is a spinor, ρ_T is the charge density and $\boldsymbol{J}$ is the current density. It is interesting that Equation (31), obtained from Dirac's equation using the differential operator bracket in Equation (30), yields a continuity equation sharing both the Dirac and Klein-Gordon features of the charge (probability) density. This interplay exists despite the fact that Dirac's equation represents a fermionic particle while Klein-Gordon equation represents a bosonic particle.

6. THE SPACE AND TIME INVARIANCE OF DIRAC'S EQUATION

If we apply the transformations in Equations (32) and (43) to Equation (29), Dirac's equation will be invariant. Thus, the space and time transformation represented by Equations (32) and (43) ushers into a new transformation of Dirac's equation that were never known before. With some scrutiny, we know from the theory of relativity that the kinetic energy (E_K) is related to the total energy (E) by $E_K = E - m_0c^2$. In quantum mechanics,

$E \to i\hbar\frac{\partial}{\partial t}$, so that

$$\hat{E}_K = i\hbar\frac{\partial}{\partial t} - m_0c^2 = i\hbar\left(\frac{\partial}{\partial t} + i\frac{m_0c^2}{\hbar}\right) = i\hbar\frac{\partial}{\partial\eta}. \tag{54}$$

This is the relativistic kinetic energy operator. Alternatively, using Equation (29), this can be written as

$$\hat{E}_K = c\boldsymbol{\alpha}\cdot\hat{\boldsymbol{p}}. \tag{55}$$

This equation implies that Dirac's equation can be obtained from the relativistic energy equation

$$E = E_K + \beta m_0 c^2. \tag{56}$$

This equation suggests that there are two possible energy equations. These are $E_{\pm} = E_K \pm m_0 c^2$. Hence, a Dirac's particle has in principle two energies, $E_+ = E_K + m_0 c^2$ and $E_- = E_K - m_0 c^2$.

Using Equation (32), Dirac's equation Equation (29), becomes

$$\frac{\partial \psi}{\partial \eta} + \nabla \cdot (c\boldsymbol{\alpha})\psi = 0. \tag{57}$$

Thus, in the time coordinate η, Dirac's equation represents a continuity-like equation. However, in the real time, the continuity equation in Dirac's formalism, is defined as

$$\frac{\partial \rho}{\partial t} + \nabla \cdot \boldsymbol{J} = 0, \tag{58}$$

where $\rho = \psi^+ \psi$ and $\boldsymbol{J} = \psi^+ c\boldsymbol{\alpha}\psi$. Using Equation (43), Dirac's equation is transformed into a continuity-like equation in the new space coordinate, viz.,

$$\frac{\partial \psi}{\partial t} + \nabla' \cdot (c\boldsymbol{\alpha})\psi = 0. \tag{59}$$

Notice here that $\boldsymbol{J}$ has the same form in both coordinates.

7. Space and Time Invariance of Maxwell's Equations

We would like here to apply the space and time transformations in Equations (32) and (43) to explore their implications in Maxwell's equations. These transformations leave Dirac's equation invariant. We know that quantum electrodynamics incorporates the interaction of an electron with a photon. Quantum electrodynamics becomes invariant under gauge transformation, if we replace the partial derivative with a covariant derivative incorporateing the photon field. Analogously, we assume here that Maxwell's equations are invariant under the new space and time transformations in Equations (32) and (43). Applying Equations (32) and (43) to the Ampere's and Faraday's equations [12].

$$\nabla \times \boldsymbol{E} = -\frac{\partial \boldsymbol{B}}{\partial t}, \tag{60}$$

and

$$\nabla \times \boldsymbol{B} = \mu_0 \boldsymbol{J} + \frac{1}{c^2}\frac{\partial \boldsymbol{E}}{\partial t}, \tag{61}$$

yield

$$\boldsymbol{B} = \frac{\boldsymbol{v} \times \boldsymbol{E}}{c^2}, \quad \boldsymbol{v} = c\boldsymbol{\alpha}, \tag{62}$$

and

$$\boldsymbol{E} = -\boldsymbol{v} \times \boldsymbol{B}, \quad \boldsymbol{v} = c\boldsymbol{\alpha}. \tag{63}$$

The remaining two Maxwell's equations

$$\nabla \cdot \boldsymbol{E} = \frac{\rho}{\varepsilon_0}, \quad \nabla \cdot \boldsymbol{B} = 0. \tag{64}$$

yield

$$\boldsymbol{v} \cdot \boldsymbol{E} = 0, \quad \boldsymbol{v} \cdot \boldsymbol{B} = 0, \quad \boldsymbol{v} = c\boldsymbol{\alpha}. \tag{65}$$

It is interesting to notice that Equation (65) is compatible with Equations (62) and (63). Moreover, Equations (62) and (63) define the relations between the electric and magnetic fields produced by the moving charge. If the electric (magnetic) field is known, one can obtain the corresponding magnetic (electric) field. Equation (63) shows that the charge moving with constant velocity experiences no net force. The electric field lines of a moving charge crowded in the direction perpendicular to $\boldsymbol{v}$ and are given by [13]

$$\boldsymbol{E} = \frac{1}{4\pi\varepsilon_0}\frac{q\left(1-\beta^2\right)}{\left(1-\beta^2\sin^2\theta\right)^{\frac{3}{2}}}\frac{\boldsymbol{r}}{r^3}, \quad \beta = \frac{\boldsymbol{v}}{c}, \tag{66}$$

and

$$B = \frac{\mu_0}{4\pi} \frac{q\left(1-\beta^2\right)}{\left(1-\beta^2 \sin^2\theta\right)^{\frac{3}{2}}} \frac{\boldsymbol{v}\times\boldsymbol{r}}{r^3}, \quad \beta = \frac{v}{c}. \tag{67}$$

Equation (66) shows that the electric and magnetic fields of a moving charge in the forward direction ($\theta = 0$) are less than the electric field of stationary charge. However, the electric and magnetic fields in the perpendicular direction ($\theta = \pi/2$) are bigger than the electric field of stationary charge. Equations (66) and (67) give the relations between the electric and magnetic fields produced by a moving charged particle with constant velocity $\boldsymbol{v}$. This velocity is given by $\boldsymbol{v} = c\boldsymbol{\alpha}$. This coincides with the quantum mechanics definition of the particle velocity [1-3]. Equation (65) shows that the electric and magnetic fields produced by the charged particle are always perpendicular to the particle's direction of motion. Equation (67) gives, at low velocity, the Biot-Savart law. The power delivered by the fermionic charged particle by its electric and magnetic fields is given by

$$P = \boldsymbol{F}\cdot\boldsymbol{v} = q\boldsymbol{v}\cdot\boldsymbol{E} + q\boldsymbol{v}\cdot\left(\boldsymbol{v}\times\boldsymbol{B}\right) = 0$$

The application of the transformations (32) and (43) in the generalized continuity Equations (6) and (8) yields

$$\boldsymbol{v}\cdot\boldsymbol{J} = -\rho c^2, \quad \boldsymbol{J} = -\rho\boldsymbol{v}, \quad \boldsymbol{v} = c\boldsymbol{\alpha}. \tag{68}$$

Since in Dirac formalism $\rho > 0$, the current ushers in a direction opposite to the velocity direction. Moreover, for a constant velocity, one has $(\nabla\rho)\times\boldsymbol{v} = 0$. Equation (68) is very interesting since it defines the charge density (scalar) in terms of the current density (vector). Accordingly, one can define the four vectors in terms of vectorial quantities only.

8. CONCLUDING REMARKS

By introducing three vanishing differential commutator brackets for spinor fields, we have derived a variant form of Dirac's and Klein-Gordon wave equations. Dirac's equation yields a modified Klein-Gordon wave equation. This equation yields directly two energy states for the particle in question.

Moreover, Dirac's equation is found to be similar to the continuity equation. We have found time and space transformations under which Dirac's and Maxwell's equations are invariant. In terms of these coordinates, Dirac and Klein-Gordon equations describe a mass-less particle. The invariance of these transformations under Maxwell's equations shows that the electric and magnetic fields produced by a moving charge are perpendicular to velocity of the particle. Hence, there is no power associated with these fields. The space and time transformations show that space and time are quantized in terms of characteristic units of T_0 and X_0. The fermionic charged particle exhibits its wave and corpuscular nature on periodic space and time basis.

9. ACKNOWLEDGEMENTS

This work is supported by the university of Khartoum research fund. We gratefully acknowledge this support. The critical and useful comments by the anonymous referees are highly acknowledged.

REFERENCES

1. J. D. Bjorken and S. D. Drell, "Relativistic Quantum Mechanics," McGraw-Hill, New York, 1964.
2. L. D. Landau and E. M. Liftshiz, "Quantum Mechanics," 3rd Edition, Pergamon Press, Oxford, 1977.
3. V. B. Berestetskii, L. P. Pitaevskii and E. M. Lifshitz, "Quantum Electrodynamics," 2nd Edition, Vol. 4, Elsevier, Amsterdam, 1982.
4. A. I. Arbab and Z. Satti, "On the Generalized Maxwell Equations and Their Prediction of Electroscalar Wave," Progress in Physics, Vol. 2, No. 8, 2009, pp. 8-13.
5. A. I. Arbab and H. M. Widatallah, "The Generalized Continuity Equations," Chinese Physics Letters, Vol. 27, No. 8, 2010, Article ID 084703.
6. A. I. Arbab and F. A. Yassein, "A New Formulation of Electromagnetism," Journal of Electromagnetic Analysis and Applications, Vol. 2, No. 8, 2010, p. 457.

7. A. I. Arbab, "A Quaternionic Quantum Mechanics," Applied Physics Research, Vol. 3, No. 2, 2011, p. 160.

8. P. G. Tait, "An Elementary Treatise on Quaternions," 2nd Edition, Cambridge University Press, Cambridge, 1873.

9. H. F. Harmuth, T. W. Barrett and B. Meffert, "Modified Maxwell Equations in Quantum Electrodynamics," World Scientific, River Edge, 2001.

10. G. Feinberg, "Possibility of Faster-Than-Light Particles," Physical Reviews, Vol. 159, No. 5, 1967, pp. 1089-1105.

11. J. Ciborowski, "Hypothesis of Tachyonic Neutrinos," Acta Physicsa Polonica B, Vol. 29, No. 1-2, 1998, pp. 113-121.

12. J. D. Jackson, "Classical Electrodynamics," 2nd Edition, Wiley, New York, 1975.

13. F. Zbigniew, "Lecture Notes in Electromagnetic Theory," University of Queensland, Brisbane, 2005.

2
CHAPTER
ON THE QUANTUM ZENO EFFECT AND TIME SERIES RELATED TO QUANTUM MEASUREMENTS

Karl-Heinz Fichtner[1], Kei Inoue[2]

[1]*School of Mathematics and Computer Science, Institute of Applied Mathematics, Friedrich-Schiller-Universität Jena, Jena, Germany*

[2]*Department of Electrical Engineering, Faculty of Engineering, Tokyo University of Science, Yamaguchi, Sanyoonoda, Japan*

ABSTRACT

Our main aim is to prove a more general version of the quantum Zeno effect. Then we discuss some examples of the quantum Zeno effect. Furthermore, we discuss a possibility that based on the quantum Zeno effect and certain experiments one could check whether, from the statistical point of view, a concrete system behaves like a quantum system. The more general version of quantum Zeno effect can be helpful to prove that the brain acts like in a quantum system. The proof of our main result is based on certain formulas describing probability distributions of time series related to quantum measurements.

Keywords: Quantum Zeno Effect; Quantum Measurements; Time Series

1. INTRODUCTION

Zeno of Elea (ca. 490 BC-ca. 430 BC) was one of the oldest Greek philosopher. Aristotle called him the inventor of the dialectic. He is best known for his paradoxes. The most famous are the so called "arguments against motion" described by Aristotle in his Physics. Let us mention Zeno's arrow paradox which states that, since an arrow in flight is not seen to move during any single instant, it cannot possibly be moving at all. What does it mean? Let $X(0)$ be the position of arrowhead at time $t_0 = 0$. Then the arrow will move. $X(t)$ denotes the position of the arrowhead at time t. Now, let $t_1 < t_2 < \cdots < t_n$ be a sequence of times. Then the observed positions of the arrowhead are represented by the sequence $X(t_0), \cdots, X(t_n)$. If the distance of times $t_{k+1} - t_k \approx 0$ is very small then the distance of the corresponding positions $X(t_{k+1}) - X(t_k) \approx 0$ is very small. According to Zeno that implies that the arrow cannot move under continuous observation. Of course we know that this is nonsense because in classical mechanics observations are of no effect on the state of the system—not so in quantum mechanics where observation, i.e., measurement, causes a change of the state. That is the basic idea of the quantum Zeno effect.

The quantum Zeno effect is a name coined by George Sudarshan and Baidyanath Mirsa of the University of Texas in 1977 in their analysis of the situation in which an unstable particle, if observed continuously, will never decay [1]. One can "freeze" the evolution of the system by measuring it frequently enough in its (known) initial state. The meaning of the term has since expanded, leading to a more technical definition in which the time evolution can be suppressed not only by measurement [2,3]. That does not mean that there exists a complete theory of the quantum Zeno effect. Even there is no precise meaning concerning the notion of "suppression of the time evolution".

An earlier theoretical exploration of this effect of measurement was published in 1974 by Degasperis et al. [4] and A. Turing described it in 1954 [5]: It is easy to show (but there was no rigorous proof!) using standard theory that if a system starts in an eigenstate of some observable, and measurements are made of that observable N times a second, then even if the state is not a stationary one, the probability that system will be in the same state after, say, one second, tends to one as N tends to infinity; that is, that continual observations will prevent motion.

The idea is contained in the early work by J. von Neumann, sometimes called the reduction postulate [6]. According to the reduction postulate, each measurement causes the wave function to "collapse" to a pure eigenstate of the measurement basis.

Usually the model of Turing is far from the reality, because in reality the state of the quantum system is usually unknown, and it is a mixed state. Furthermore that kind of measurements is not realistic, i.e., real measuring instruments cannot perform such kind of measurements. Nevertheless a general theory of the quantum Zeno effect should contain the Turing model as a special case.

Following A. Turing the reduction of the states seems to be the background of the quantum Zeno effect. Now our work on the quantum Zeno effect was motivated by the quantum model of certain brain activities developed in [7-15]. For that reason let us briefly mention what kind of reduction of states were considered in certain quantum models of brain activities.

In Wikipedia, the free encyclopedia on the Zeno effect is written the following:

1.1. Significance to Cognitive Science

The quantum Zeno effect is becoming a central concept in exploration of controversial theories of quantum mind consciousness within the discipline of cognitive science. In his book Mindful Universe (2007) (see also [16]), Henry Stapp claims that the mind holds the brain in a superposition of states using the quantum Zeno effect.

Now, let us consider the process of recognition of signals in the brain. Then the basic idea of Stapp may be described as follows: If there will be a signal arising from the senses, the process of recognition starts. That process represents a rapid sequence of "trials" checking whether the signal from the senses and the signal created by the brain partially coincide. He identifies these trials with measurements performed by a mysterious "observer" in the brain. That does not seem to be realistic.

Related to the model of recognition discussed in [7-15], these trials are not represented by a process of measurements. Nevertheless the results of the trials are represented by projections like in the case of measurements.

That R. Penrose [17,18] explains as follows: The conventional quantum theory view is that the quantum state reduces by measurement or observation (subjective reduction). Now, a number of physicists have argued in support of special models in which the rules of standard quantum mechanics are modified by the inclusion of some additional procedure according to which the reduction of the state becomes an objectively real process (objective reduction)—the system abruptly self-collapses. Consciousness, it is argued, requires non-computability. The only readily available apparent source of non-computability is self-collapse. The self-collapse, irreversible in time, creates an instantaneous "now" event. Sequences of such events create a flow of time, and consciousness.

That means recognition of signals is a process of self-collapses. That coincides with the basic idea of the quantum model of recognition of signals in the brain in [7-15]. Furthermore, like sequences of measurements that process suppresses a certain unitary time evolution in the brain. A lot of experiments show that there is a suppression of a certain process in the brain caused by the recognition of signals or some other kind of activities of the brain (cf. [19]).

Using series of measurements at different times one tries to get information on the statistical behaviour of certain random processes. That series of measurements give a so called time series.

1.2. Time Series—Classical Probability Theory

In classical stochastics one identifies the time evolution of a random system with a stochastic process $(\xi_t)_{t\geq 0}$ taking on values in a certain space G. The random variable ξ_t represents the system at time t. A measurement at a certain time $t \geq 0$ one identifies with a certain (measurable) real function f on G, and the random number $f(\xi_t)$ represents the output of the measurement.

Now, let us consider a sequence of time $t_0 = 0 \leq t_1 < t_2 < \cdots < t_n$. At time t_k one performs a measurement related to a function f_k on G. Then the sequence $\eta_k := f_k\left(\xi_{t_k}\right)(k = 1, \cdots, n)$ represents a so called time series.

If we would know the distribution of the underlying stochastic process, then we can calculate the distribution of the time series.

Sometimes specialists use the notion "time series" only in the special case:

$$\tau = t_k - t_{k-1}; f_k = f \quad (k = 1, \cdots, n)$$

That means one performs always the same measurements at equidistant times.

In order to obtain information concerning the state of the brain and its time evolution one has to perform several EEG measurements at different times. That gives a time series. In practice specialists use certain statistical methods evaluating that time series, i.e., they analyze the outcomes of the sequence of measurements. To make these statistical investigations effective, and precise from mathematical point of view one should know the type of the distribution of the stochastic process on the space of point configurations representing the time evolution of the configuration of excited neurons in the brain. Up to now that information is lacking (cf. [20]).

1.3. The Quantum Case of Time Series

Let $(V_t)_{t\geq 0}$ be a semigroup of unitary operators on a separable Hilbert space H representing the time evolution of a quantum system. ρ denotes the state of the quantum system at time $t_0 = 0$. We consider again a sequence of time $t_0 \leq t_1 < \cdots < t_n$ and perform measurements at time $t_k\,(k = 1,\cdots,n)$ related to selfadjoint operators B_k given by their unique representation $B_k = \sum_{j=1}^{m_k} z_{k,j} \Pr_{H_{k,j}}$, i.e., $(H_{k,j})_{j=1}^{m_k}$ represents an orthogonal decomposition of H and $z_{k,j} \neq z_{k,s}$ if $j \neq s$. Because each realistic measuring instrument has a finite scale we consider only selfadjoint operators being of this type.

Formally we can identify the random outputs of these measurements with random numbers $\eta_1,\cdots,\eta_n$ representing the time series in the considered quantum case.

Unfortunately, in the quantum case there does not exist a stochastic process $(\xi_t)_{t\geq 0}$ such that we can always represent the distribution of the time series as follows:

$$P_{\eta_1,\cdots,\eta_n} = P_{f_1(\xi_{t_1}),\cdots,f_n(\xi_{t_n})}.$$

For that reason one has to develop a useful concept in order to describe the distribution of such time series related to quantum measurements. That will be done in Section 2 of this paper.

1.4. On the Quantum Zeno Effect

Starting with the notations used above we will specialize

$$B_k = B := \sum_{j=1}^{m} z_j \operatorname{Pr}_{H_j}; \quad t_k := \frac{k-1}{n} T \quad (k = 1, \cdots, n).$$

That means we consider repeated measurements in the time interval $[0, T[$ according to B starting at time $t_1 = t_0 = 0$ with equidistant times. Again ρ denotes the state of the quantum system at time 0, and the random numbers $\eta_1, \cdots, \eta_n$ represent the outputs of the measurements.

In Section 3 of this paper we will prove the following theorem.

Theorem It holds

$$P(\eta_k = \eta_1; k = 2, \cdots, n) \xrightarrow[n \to +\infty]{} 1.$$

Immediately from the theorem we obtain the following corollary.

Corollary It holds

$$P(\eta_k = z_j; k = 1, 2, \cdots, n)$$

$$\xrightarrow[n \to +\infty]{} P(\eta_1 = z_j) = tr\left(\rho \operatorname{Pr}_{H_j}\right), \quad j = 1, \cdots, m.$$

That does not mean that the state will not be changed in time. This is related to the fact that using only one kind of measurement one cannot determine the state. One can consider a simple example illustrating that. Nevertheless in some special cases one has a total suppression of the time evolution in this sense that the state is not changing.

1.5. Example—A. Turing

Remember the statement of A. Turing: If a system starts in an eigenstate of some observable, and measurements are made of that observable N times a second, then the probability that the system will be in the same state after, say, one second, tends to one as N tends to infinity.

That we obtain specializing as follows: Let $\rho = \mathrm{Pr}_\Psi$ be a state related to a normalized $\Psi \in H$, and let the measurements be related to the selfadjoint operator $B = b\,\mathrm{Pr}_\Psi + \hat{B}$, where $\hat{B}$ is a selfadjoint operator such that $\hat{B}$ and Pr_Ψ are orthogonal, and the eigenvalues of $\hat{B}$ are different from b. Then $\rho = \mathrm{Pr}_\Psi$ represents an eigenstate of the operator B related to the eigenvalue b.

Now, one has to perform measurements according to B at the times

$$t_k = \frac{k-1}{N} T \quad (k = 1, \cdots, N+1).$$

Then we have $t_1 = 0$, $t_{N+1} = T$. Again η_k denotes the random output of the measurement at time $t_k \; (k = 1, \cdots, N+1)$.

Then using the corollary mentioned above we obtain

$$P(\eta_k = b; k = 1, \cdots, N+1) \xrightarrow[n \to +\infty]{} P(\eta_1 = b).$$

Because ρ is an eigenstate of B related to the eigenvalue b we have

$$P(\eta_1 = b) = 1.$$

Thus we can conclude

$$P(\eta_k = b; k = 1, \cdots, N+1) \xrightarrow[n \to +\infty]{} 1.$$

That implies

$$P(\eta_{N+1} = b) \xrightarrow[n \to +\infty]{} 1.$$

Because of $t_{N+1} = T$ the random value η_{N+1} would represent the output of a measurement at time T and the event "$\eta_{N+1} = b$" means that at time T the system would be in the initial state Pr_Ψ. Thus we can conclude that the probability that at time T the system will be in the same state as at the beginning tends to one as N tends to infinity.

1.6. Example—Arrow Paradox

In quantum mechanics one uses the Hilbert space $H = L^2(R^3)$ in order to describe a quantum particle located in the space R^3. Now, again let $\rho = \mathrm{Pr}_\Psi$ be a pure state related to a normalized wave function $\Psi \in H$. We interpret ρ as the state of the arrowhead at time $t_0 = 0$, and we assume that the wave function Ψ is concentrated on a "very small region" D, thus the position of the arrowhead at time 0 is "almost fixed".

Now, let H_D be the space of the wave functions concentrated on D. We specialize $B := \mathrm{Pr}_{H_D}$. Then the corresponding measurement means: "checking whether the arrowhead will be inside of the region D or not". Consequently the event "$\eta_k = 1$" means that "at time t_k the arrowhead is located inside of the region D".

Because of $P(\eta_1 = 1) = 1$ the corollary mentioned above may be rewritten as follows

$$P(\text{“at time } t_k \text{ the arrowhead is located in } D(k = 1, \cdots, n)\text{”}) \xrightarrow[n \to +\infty]{} 1$$

i.e., the arrow will not move under "continual observation".

1.7. Can We Use the Quantum Zeno Effect in Order to "Prove" That the Brain Acts Like a Quantum System?

More and more specialists in life science are convinced that one should use quantum models in order to describe some special bio-systems, others are against quantum models. Now we want to propose a certain experiment that may be helpful to decide whether a concrete system behaves like a quantum system or not.

The basic idea is as follows: First perform a sequence of measurements with very high frequency, i.e., the length n of the time series $\eta_1, \cdots, \eta_n$ obtained within, say, one second is very large. Then produce a time series $\hat{\eta}_1, \cdots, \hat{\eta}_m$ where m is very small (e.g. $m = 2$). Now, compare the correlations $cor(\eta_1, \eta_n)$ and $cor(\hat{\eta}_1, \hat{\eta}_m)$. In the idealistic quantum case it would be $cor(\eta_1, \eta_n) \approx 1$ because of the quantum Zeno effect. But that is not realistic because of certain classical noise caused by the measuring procedure. Nevertheless, in the quantum case $|cor(\hat{\eta}_1, \hat{\eta}_m)|$ should be significant lower than $|cor(\eta_1, \eta_n)|$—which would contradict to the classical case.

Finally let us mention that some parts of this paper are based on [21].

2. DISTRIBUTION OF A TIME SERIES RELATED TO QUANTUM MEASUREMENTS

Let $(V_t)_{t \geq 0}$ be a semigroup of unitary operators on a separable Hilbert space H representing the time evolution of a quantum system. ρ denotes the state of the quantum system at time $t_0 = 0$.

We consider a sequence of times $t_0 \leq t_1 < \cdots < t_n$ and perform measurements at time $t_k \left(k = 1, \cdots, n\right)$ related to selfadjoint operators B_k given by their unique representation

$$B_k := \sum_{j=2}^{m_k} z_{k,j} \, \mathrm{Pr}_{H_{k,j}}, \quad k = 1,\cdots,n.$$

i.e., $(H_{k,j})_{j=1}^{m_k}$ represents an orthogonal decomposition of H and $z_{k,j} \neq z_{k,s}$ if $j \neq s \ (k = 1,\cdots,n)$. Formally the random outputs of these measurements we identify with random numbers $\eta_1,\cdots,\eta_n$.

Theorem 1 Let ρ be a pure state related to $x \in H$, $\|x\| = 1$. Further we put $\tau_k := t_k - t_{k-1} \ (k = 1,\cdots,n)$. Then we get the probabilities

$$\begin{aligned} & P\left(\eta_1 = z_{1,j_1},\cdots,\eta_n = z_{n,j_n}\right) \\ & = \left\| \left(\mathrm{Pr}_{H_{n,j_n}} V_{\tau_n}\right)\cdots\left(\mathrm{Pr}_{H_{1,j_1}} V_{\tau_1}\right)x \right\|^2, \\ & j_1 = 1,\cdots,m_1,\cdots,j_n = 1,\cdots,m_n. \end{aligned}$$

Proof 1 We put

$$\begin{aligned} & p_k\left(z_{1,j_1},\ldots,z_{k,j_k}\right) \\ & := \left\| \left(\mathrm{Pr}_{H_{k,j_k}} V_{\tau_k}\right)\cdots\left(\mathrm{Pr}_{H_{1,j_1}} V_{\tau_1}\right)x \right\|^2, \\ & j_1 = 1,\cdots,m_1,\cdots,j_k = 1,\cdots,m_k; k = 1,\cdots,n. \end{aligned} \tag{2.1}$$

Then we get

$$P\left(\eta_1 = z_{1,j_1}\right) = p_1\left(z_{1,j_1}\right), \quad j_1 = 1,\cdots,m_1. \tag{2.2}$$

Now let us assume that for some k it holds

$$\begin{aligned} & P\left(\eta_1 = z_{1,j_1},\cdots,\eta_k = z_{k,j_k}\right) \\ & = p_k\left(z_{1,j_1},\cdots,\eta_k = z_{k,j_k}\right), \\ & j_1 = 1,\cdots,m_1,\cdots,j_k = 1,\cdots,m_k. \end{aligned} \tag{2.3}$$

Using J. von Neumann's rule we can conclude that under condition „$\eta_1 = z_{1,j_1},\cdots,\eta_k = z_{k,j_k}$" (i.e., knowing the exact results of the

k measurements) immediately after the k -th measurements the quantum system is in the pure state

$$\frac{\left(\Pr_{H_{k,j_k}} V_{\tau_k}\right)\ldots\left(\Pr_{H_{1,j_1}} V_{\tau_1}\right)x}{\left\|\left(\Pr_{H_{k,j_k}} V_{\tau_k}\right)\ldots\left(\Pr_{H_{1,j_1}} V_{\tau_1}\right)x\right\|}.$$

Then at time t_{k+1} the system is in the state

$$\frac{V_{\tau_{k+1}}\left(\Pr_{H_{k,j_k}} V_{\tau_k}\right)\ldots\left(\Pr_{H_{1,j_1}} V_{\tau_1}\right)x}{\left\|\left(\Pr_{H_{k,j_k}} V_{\tau_k}\right)\ldots\left(\Pr_{H_{1,j_1}} V_{\tau_1}\right)x\right\|}.$$

Now performing the measurement according to B_{k+1} we get the value $z_{k+1,j_{k+1}}$ with (conditional) probability

$$P\left(\eta_{k+1} = z_{k+1,j_{k+1}} \middle| \eta_1 = z_{1,j_1}, \cdots, \eta_k = z_{k,j_k}\right)$$
$$= \frac{\left\|\left(\Pr_{H_{k+1,j_{k+1}}} V_{\tau_{k+1}}\right)\ldots\left(\Pr_{H_{1,j_1}} V_{\tau_1}\right)x\right\|^2}{\left\|\left(\Pr_{H_{k,j_k}} V_{\tau_k}\right)\ldots\left(\Pr_{H_{1,j_1}} V_{\tau_1}\right)x\right\|^2}. \tag{2.4}$$

Using (2.1), (2.3) and (2.4) we obtain

$$P\left(\eta_1 = z_{1,j_1}, \cdots, \eta_{k+1,j_{k+1}}\right) = p_{k+1}\left(z_{1,j_1}, \cdots, z_{k+1,j_{k+1}}\right),$$
$$j_1 = 1, \cdots, m_1, \cdots, j_{k+1} = 1, \cdots, m_{k+1}.$$

That proves Theorem 1.

Now let $f_1, \cdots, f_n$ be measurable functions. We will consider the expectation

$$E\prod_{r=1}^{n} f_r\left(\eta_r\right) = \sum_{j_1, \cdots, j_n} p_n\left(z_{1,j_1}, \cdots, z_{n,j_n}\right)\prod_{r=1}^{n} f_r\left(z_{r,j_r}\right). \tag{2.5}$$

Example 1 Let $D_1,\cdots,D_n$ be measurable subsets of R. We consider the indicator functions $f_k=\chi_{D_k}\,(k=1,\cdots,n)$. Then we have

$$E\prod_{r=1}^{n}f_r(\eta_r)=P(\eta_1\in D_1,\cdots,\eta_n\in D_n).$$

Example 2 We consider the case $f_k(y)=y^s$, $(r\neq k)\ f_r\equiv 1$. Then we have

$$E\prod_{r=1}^{n}f_r(\eta_r)=E\eta_k^s.$$

Example 3 We consider the case $f_k(y)=f_j(y)=y,\ f_r\equiv 1(j\neq r\neq k)$. Then we have

$$E\prod_{r=1}^{n}f_r(\eta_r)=E\eta_k\eta_j.$$

Using Example 2 and Example 3 we can calculate correlations.

Example 4 We consider the case

$$f_r(y)=\mathrm{e}^{is_ry}\,(r=1,\cdots,n).$$

Then

$$\gamma(s_1,\cdots,s_n)=E\prod\nolimits_{r=1}^{n}f_r(\eta_r)$$

represents the characteristic function of the vector $[\eta_1,\cdots,\eta_n]$. Observe that these characteristic functions represent the complete information concerning the distribution of time series.

Now in quantum mechanics expectations are of the type $tr(\rho A)$ where ρ represents the state of the system and A is a certain (selfadjoint) operator. For that reason we will construct an operator A related to $B_1,t_1,f_1,\cdots,B_n,t_n,f_n$ such that we obtain

$$tr(\rho A)=E\prod_{r=1}^{n}f_r(\eta_r). \tag{2.6}$$

For that purpose we define the following linear mappings from the set $\hat{B}(H)$ of bounded operators on H into this set putting

$$K_t(B) := V_t^* B V_t, \quad B \in \hat{B}(H), t \geq 0, \tag{2.7}$$

$$W_{B_k, f_k}(B) := \sum_{j=1}^{m_k} f_k(z_{k,j}) \Pr_{H_{k,j}} B \Pr_{H_{k,j}},$$
$$B \in \hat{B}(H); k = 1, \cdots, n. \tag{2.8}$$

Then we put

$$A := A_{B_1, t_1, f_1, \cdots, B_n, t_n, f_n}$$
$$:= K_{\tau_1} \circ W_{B_1, f_1} \circ \cdots \circ K_{\tau_{n-1}} \circ W_{B_{n-1}, f_{n-1}} \left(K_{\tau_n}\left(f_n(B_n)\right)\right). \tag{2.9}$$

If the operator B is selfadjoint then $K_t(B)$ is selfadjoint. Further one easily checks that $W_{B_k, f_k}(B)$ is selfadjoint if B is selfadjoint and f_k is a real function. For that reason we can conclude that the operator A is selfadjoint if $f_k(k = 1, \cdots, n)$ are real functions.

Immediately from the definition of $A_{B_1, t_1, f_1, \cdots, B_n, t_n, f_n}$ we get

$$A_{B_1, t_1, f_1, \cdots, B_n, t_n, f_n} = K_{t_1}\left(A_{B_1, 0, f_1, B_2, t_2 - t_1, f_2, \cdots, B_n, t_n - t_1, f_n}\right). \tag{2.10}$$

Finally we obtain the following theorem.

Theorem 2 Let the operator A be given by (2.9). Then (2.6) holds.

Proof 2 We can restrict ourselves to the case of a pure state related to $x \in H$, $\|x\| = 1$. Then we have to prove that it holds

$$\langle x | Ax \rangle = E \prod_{r=1}^{n} f_r(\eta_r). \tag{2.11}$$

Using Theorem 1 and (2.5) we get

$$E\prod_{r=1}^{n} f_r(\eta_r) = \sum_{j_1,\cdots,j_n}\left(\prod_{r=1}^{n} f_r\left(z_{r,j_r}\right)\right) \cdot \left\|\left(\Pr_{H_{n,j_n}} V_{\tau_n}\right)\cdots\left(\Pr_{H_{1,j_1}} V_{\tau_1}\right)x\right\|^2. \tag{2.12}$$

Further we have

$$\begin{aligned}
&\left\|\left(\Pr_{H_{n,j_n}} V_{\tau_n}\right)\cdots\left(\Pr_{H_{1,j_1}} V_{\tau_1}\right)x\right\|^2 \\
&= \left\langle x \middle| \left(\Pr_{H_{1,j_1}} V_{\tau_1}\right)^* \cdots \left(\Pr_{H_{n,j_n}} V_{\tau_n}\right)^* \right. \\
&\qquad \left. \left(\Pr_{H_{n,j_n}} V_{\tau_n}\right)\cdots\left(\Pr_{H_{1,j_1}} V_{\tau_1}\right)x\right\rangle \\
&= \left\langle x \middle| V_{\tau_1}^* \Pr_{H_{1,j_1}} \cdots V_{\tau_{n-1}}^* \Pr_{H_{n-1,j_{n-1}}} V_{\tau_n}^* \Pr_{H_{n,j_n}} \right. \\
&\qquad \left. V_{\tau_n} \Pr_{H_{n-1,j_{n-1}}} V_{\tau_{n-1}} \cdots \Pr_{H_{1,j_1}} V_{\tau_1} x\right\rangle.
\end{aligned} \tag{2.13}$$

Finally using (2.13), (2.12), (2.9), (2.8) and (2.7) we obtain (2.11). That proves theorem 2.

Remark 1 The time evolution of a quantum mechanical system usually is characterized by a semigroup $(V_t)_{t\geq 0}$ of unitary operators. Then the state ρ_t of the system at time t is characterized by

$$tr(\rho_t B) = tr(\rho K_t(B)), \quad B \in \hat{B}(H), t \geq 0. \tag{2.14}$$

If in addition to that time evolution there are measurements like in the case we have considered this "perturbed" evolution in general no longer may be described by a semigroup of unitary operators. Though the ansatz (2.14) still makes sense. One has to replace the semigroup $(K_t)_{t\geq 0}$ by a family of completely positive linear maps $\hat{K}_t$ on $\hat{B}(H)$. In the case considered above these maps are defined as follows:

$$\hat{K}_t(B) := \begin{cases} K_t(B), & 0 \leq t < t_1, \\ K_{\tau_1} \circ W_{B_1,1} \circ \cdots \circ K_{\tau_m} \circ W_{B_m,1}\left(K_{t-\tau_m}(B)\right), & t_m \leq t < t_{m+1}, \\ K_{\tau_1} \circ W_{B_1,1} \circ \cdots \circ K_{\tau_n} \circ W_{B_n,1}\left(K_{t-\tau_n}(B)\right), & t_n \leq t, \end{cases} \quad B \in \hat{B}(H). \tag{2.15}$$

Further let us mention that one can generalize our considerations replacing from the beginning the semigroup $(K_t)_{t\geq 0}$ by an arbitrary semigroup of identity preserving completely positive linear maps on $\hat{B}(H)$.

Remark 2 Let us consider the trivial case of a time evolution

$$K_t(B) = B, \quad B \in \hat{B}(H). \tag{2.16}$$

Further we consider the case of only two measurements, i.e., $n = 2$. Then we obtain the probabilities

$$P_{B_1,B_2}(z_{1,j_1}, z_{2,j_2}) = \left\| \Pr_{H_{2,j_2}} \Pr_{H_{1,j_1}} x \right\|^2, \quad j_1 = 1,\cdots,m_1; j_2 = 1,\cdots,m_2. \tag{2.17}$$

The procedure may be interpreted as follows: first perform a measurement according to B_1 and then perform a measurement according to B_2.

Now we change that order, i.e., first perform a measurement according to B_2 and then perform a measurement according to B_1. Then we obtain the probabilities

$$P_{B_2,B_1}(z_{2,j_2}, z_{1,j_1}) = \left\| \Pr_{H_{1,j_1}} \Pr_{H_{2,j_2}} x \right\|^2, \quad j_1 = 1,\cdots,m_1; j_2 = 1,\cdots,m_2. \tag{2.18}$$

If B_1 and B_2 commute then we get like in the classical case

$$P_{B_1,B_2}(z_{1,j_1}, z_{2,j_2}) = P_{B_2,B_1}(z_{2,j_2}, z_{1,j_1}), j_1 = 1,\cdots,m_1; j_2 = 1,\cdots,m_2. \tag{2.19}$$

In that case specialists say (cf. [22,23]) that "there exist joint probabilities".

If B_1 and B_2 do not commute one cannot expect that (2.19) holds in general, i.e., for all states. That situation is related to what specialists call "breaking the classical probability law".

In the following some comments: From (2.17) and (2.18) we obtain two probability measures P_{B_1,B_2} and P_{B_2,B_1} on R^2 related to two random vectors each of them consisting of two components. In both cases we can define the following (classical) conditional probabilities

(a) "distribution of second component under condition we know the first component"

(b) "distribution of first component under condition we know the second component"

The conditional distributions (a) makes sense from the practical point of view. It represents the behaviour of the output of the second measurement under condition that we know the output of the first measurement. The conditional distribution of the first measurement under condition that we know the output of the second measurement (case (b)) may be senseless because the first measurement has already been done.

Furthermore, one has to take into account that P_{B_1,B_2} and P_{B_2,B_1} describe two different procedures. For that reason it seems to be obvious that using conditional probabilities calculated according to P_{B_1,B_2} for calculations based on P_{B_2,B_1} will cause some trouble, i.e., basic formulas from classical probability will be violated if $P_{B_1,B_2} \neq P_{B_2,B_1}$.

Nevertheless, from the mathematical point of view it is interesting to check whether one has joint probabilities because in that case one can apply formulas from classical probability like Bayes formula.

3. ON THE QUANTUM ZENO EFFECT

Starting with the notations used in the previous section we will specialize

$$B_k = B := \sum_{j=1}^{m} z_j \operatorname{Pr}_{H_j}, \quad k = 1, \cdots, n. \tag{3.1}$$

$$t_k := \frac{k-1}{n} T, \quad k = 1, \cdots, n. \tag{3.2}$$

That means we consider repeated measurements in the time interval $[0,T[$ according to B starting at time $t_1 = t_0 = 0$ with distance of time

$$\tau_k = \tau := \frac{1}{n} T, \quad k = 2, \cdots, n. \tag{3.3}$$

Further we consider the following semigroup of unitary operators on H

$$V_t := \mathrm{e}^{-ith}, \quad t \geq 0 \tag{3.4}$$

where the selfadjoint operator h is given by its unique representation

$$h := \sum_{j=1}^{r} c_j \, \mathrm{Pr}_{D_j} \tag{3.5}$$

i.e., $(D_j)_{j=1}^{r}$ represents an orthogonal decomposition of

H and $c_j \neq c_s$ if $j \neq s$.

Again ρ denotes the state of the quantum system at time 0, and the random numbers $\eta_1, \cdots, \eta_n$ represent the outputs of measurements.

Then we obtain the following theorem.

Theorem 3 It holds

$$P(\eta_k = \eta_1; k = 2, \cdots, n) \xrightarrow[n \to +\infty]{} 1. \tag{3.6}$$

Immediately from theorem 3 we get the following corollary.

Corollary 1

$$\begin{aligned} &P(\eta_k = z_j; k = 1, 2, \cdots, n) \xrightarrow[n \to +\infty]{} P(\eta_1 = z_j) \\ &= tr\left(\rho \, \mathrm{Pr}_{H_j}\right), \ j = 1, \cdots, m. \end{aligned} \tag{3.7}$$

Proof of Theorem 3:

We have

$$P(\eta_k = \eta_1; k = 2, \cdots, n) = 1 - \sum_{k=2}^{n} P(\eta_k \neq \eta_1, \eta_{k-1} = \eta_1, \cdots, \eta_2 = \eta_1) = 1 - \sum_{k=2}^{n} \sum_{j=1}^{m} P(\eta_k \neq z_j, \eta_{k-1} = z_j, \cdots, \eta_1 = z_j). \tag{3.8}$$

For that reason (3.6) is equivalent to

$$\sum_{k=2}^{n} P\left(\eta_k \neq z_j, \eta_s = z_j; s = 1, \cdots, k-1\right) \xrightarrow[n \to +\infty]{} 0,$$
$$j = 1, \cdots, m. \tag{3.9}$$

Now we fix $k(=2,\cdots,n)$, $j(=1,\cdots,m)$, and define indicator functions putting

$$f_s := \begin{cases} \chi_{\{z_j\}}, & s = 1, \cdots, k-1, \\ \chi_{\{y \in R: y \neq z_j\}}, & s = k, \\ 1, & s = k+1, \cdots, n. \end{cases} \tag{3.10}$$

Then we get (cf. (2.9))

$$A_{B,t_1,f_1,\cdots,B,t_n,f_n}$$
$$= V_\tau^* \left[\left(\left(V_\tau \Pr_{H_j} \right)^{k-1} \right)^* \left(I_H - \Pr_{H_j} \right) \left(V_\tau \Pr_{H_j} \right)^{k-1} \right] V_\tau. \tag{3.11}$$

Further because of Theorem 2 we have

$$P\left(\eta_k \neq z_j, \eta_{k-1} = z_j, \cdots, \eta_1 = z_j\right)$$
$$[0,T[= tr\left(\rho A_{B,t_1,f_1,\cdots,B,t_n,f_n}\right). \tag{3.12}$$

Proving Theorem 3, we can restrict ourselves to the case of a pure state ρ related to $x \in H$, $\|x\| = 1$.

Then using (3.11) and (3.12) we obtain

$$P\left(\eta_k \neq z_j, \eta_{k-1} = z_j, \cdots, \eta_1 = z_j\right) = \left\langle \left(V_\tau \Pr_{H_j}\right)^{k-1} V_\tau x \middle| \left(I_H - \Pr_{H_j}\right)\left(V_\tau \Pr_{H_j}\right)^{k-1} V_\tau x \right\rangle$$
$$= \left\langle \left(V_\tau \Pr_{H_j}\right)^{k-2} V_\tau x \middle| \Pr_{H_j} V_\tau^* \left(I_H - \Pr_{H_j}\right) V_\tau \Pr_{H_j} \left(V_\tau \Pr_{H_j}\right)^{k-2} V_\tau x \right\rangle. \tag{3.13}$$

Using (3.4) and (3.5) we get

$$V_\tau^* \left(I_H - \Pr_{H_j}\right) V_\tau = I_H - V_\tau^* \Pr_{H_j} V_\tau = \sum_{s_1, s_2}^{r} \left(1 - e^{i\tau\left(c_{s_1} - c_{s_2}\right)}\right) \Pr_{D_{s_1}} \Pr_{H_j} \Pr_{D_{s_2}} + I_H - \Pr_{H_j}.$$

That implies

$$\Pr_{H_j} V\tau^*\left(I_H - \Pr_{H_j}\right)V_\tau \Pr_{H_j} = \Pr_{H_j}\left[\sum_{s_1,s_2}^{r}\left(1-e^{i\tau\left(c_{s_1}-c_{s_2}\right)}\right)\Pr_{D_{s_1}}\Pr_{H_j}\Pr_{D_{s_2}}\right]\Pr_{H_j}$$
$$= -\Pr_{H_j}\left[\sum_{s_1,s_2}^{r}\left(\sum_{l\geq 2}\frac{\left[i\tau\left(c_{s_1}-c_{s_2}\right)\right]^l}{l!}\right)\Pr_{D_{s_1}}\Pr_{H_j}\Pr_{D_{s_2}}\right]\Pr_{H_j}$$
$$-\Pr_{H_j}\left[\sum_{s_1,s_2}^{r} i\tau\left(c_{s_1}-c_{s_2}\right)\Pr_{D_{s_1}}\Pr_{H_j}\Pr_{D_{s_2}}\right]\Pr_{H_j}. \tag{3.14}$$

Easy calculations show

$$\Pr_{H_j}\left[\sum_{s_1,s_2}^{r} i\tau\left(c_{s_1}-c_{s_2}\right)\Pr_{D_{s_1}}\Pr_{H_j}\Pr_{D_{s_2}}\right]\Pr_{H_j} = i\tau\left(\Pr_{H_j} h \Pr_{H_j} I_H \Pr_{H_j} - \Pr_{H_j} I_H \Pr_{H_j} h \Pr_{H_j}\right) = 0. \tag{3.15}$$

Because of $\|x\| = 1$ we obtain using the Schwartz inequality

$$\left|\left\langle\left(V_\tau \Pr_{H_j}\right)^{k-2} V_\tau x \middle| \Pr_{H_j}\Pr_{D_{s_1}}\Pr_{H_j}\Pr_{D_{s_2}}\Pr_{H_j}\left(V_\tau \Pr_{H_j}\right)^{k-2} V_\tau x\right\rangle\right| \leq 1. \tag{3.16}$$

Now using (3.3), (3.13), (3.14), (3.15) and (3.16) we can estimate

$$\sum_{k=2}^{n} P\left(\eta_k \neq z_j, \eta_s = z_j; s = 1,\cdots,k-1\right) \leq \frac{T^2}{n}\sum_{s_1,s_2}^{r}\left(c_{s_1}-c_{s_2}\right)^2 e^{\frac{T}{n}\left|c_{s_1}-c_{s_2}\right|},\quad j = 1,\cdots,m. \tag{3.17}$$

That implies (3.9) what proves Theorem 3.

Remark 3 Using (3.8) and (3.17) we obtain

$$1 - P\left(\eta_k = \eta_1; k = 2,\cdots,n\right)$$
$$\leq \frac{T^2 m}{n}\sum_{s_1,s_2}^{r}\left(c_{s_1}-c_{s_2}\right)^2 e^{\frac{T}{n}\left|c_{s_1}-c_{s_2}\right|}.$$

Thus for given $\varepsilon > 0$ we can calculate n_ε such that always holds

$$P\left(\eta_k = \eta_1; k = 2,\cdots,n\right) \geq 1 - \varepsilon \quad \left(n \geq n_\varepsilon\right).$$

REFERENCES

1. E. C. G. Sudarshan and B. Misra, "The Zeno's Paradox in Quantum Theory," Journal of Mathematical Physics, Vol. 18, No. 4, 1977, pp. 756-763.

2. T. Nakanishi, K. Yamane and M. Kitano, "AbsorptionFree Optical Control of Spin Systems: The Quantum Zeno Effect in Optical Pumping," Physical Review A, Vol. 65, No. 1, 2001, Article ID: 013404.

3. P. Facchi, D. A. Lidar and S. Pascazio, "Unification of Dynamical Decoupling and the Quantum Zeno Effect," Physical Review A, Vol. 69, No. 3, 2004, Article ID: 032314.

4. A. Degasperis, L. Fonda and G. C. Ghirardi, "Does the Lifetime of an Unstable System Depend on the Measuring Apparatus?" Il Nuovo Cimento A Series 11, Vol. 21, No. 3, 1974, pp. 471-484.

5. C. Teuscher and D. Hofstadter, "Alan Turing: Life and Legacy of a Great Thinker," Springer, Berlin, Heidelberg, 2004.

6. J. von Neumann, "Mathematische Grundlagen der Quantenmechanik," Springer, Berlin, Heidelberg, 1932.

7. K.-H. Fichtner and L. Fichtner, "Bosons and a quantum model of the brain, Jenaer Schriften zur Mathematik und Informatik," Math/Inf/08/05, Faculty of Mathematics and Informatics, FSU Jena, Jena, 2005, 27 p.

8. K.-H. Fichtner and L. Fichtner, "Quantum Models of Brain Activities I—Recognition of Signals," In: J. C. Garcia, R. Quezada and S. B. Sontz, Eds., Quantum Probability and Related topics, XXIII of QP-PQ: Quantum Probability and White Noise Analysis, World Scientific, New Jersey, London, Singapore, 2008, pp. 135-144.

9. K.-H. Fichtner, L. Fichtner, W. Freudenberg and M. Ohya, "On a Mathematical Model of Brain Activities," Quantum Theory, Reconsideration of Foundations-4, 962 of AIP Conference Proceedings, American Institute of Physics, Melville, New York, 2007, pp. 85-90.

10. K.-H. Fichtner, L. Fichtner, W. Freudenberg and M. Ohya, "On a Quantum Model of the Recognition Process," In: L. Accardi, W. Freudenberg and M. Ohya, Eds., Quantum Bio-Informatics, XXI of QP-PQ: Quantum Probability and White Noise Analysis, World Scientific, New Jersey, London, Singapore, 2008, pp. 64-84.

11. K.-H. Fichtner, L. Fichtner, W. Freudenberg and M. Ohya, "On a Quantum Model of the Brain Activities," In: L. Accardi, W. Freudenberg and M. Ohya, Eds., Quantum Bio-Informatics III, XXVI of QP-PQ: Quantum Probability and White Noise Analysis, World Scientific, New Jersey, London, Singapore, 2010, pp. 81-92.

12. K.-H. Fichtner, L. Fichtner, W. Freudenberg and M. Ohya, "Quantum Models of the Recognition Process—On a Convergence Theorem," Open Systems and Information Dynamics, Vol. 17, No. 2, 2010, pp. 161-187.

13. K.-H. Fichtner, L. Fichtner, W. Freudenberg and M. Ohya, "Self-Collapses of Quantum Systems and Brain Activities," In: L. Accardi, W. Freudenberg and M. Ohya, Eds., Quantum Bio-Informatics IV, XXVIII of QP-PQ: Quantum Probability and White Noise Analysis, World Scientific, New Jersey, London, Singapore, 2011, pp. 101-115.

14. K.-H. Fichtner, L. Fichtner, K. Inoue and M. Ohya, "Internal Noise Caused by the Memory," Open Systems and Information Dynamics, Vol. 18, No. 4, 2011, pp. 405- 422.

15. K.-H. Fichtner, K. Inoue and M. Ohya, "On a Quantum Model of Brain Activities—Distribution of the Outcomes of EEG-Measurements and Random Point Fields," Open Systems and Information Dynamics, Vol. 19, No. 4, 2012, Article ID: 1250025.

16. J. M. Schwartz, H. P. Stapp and M. Beauregard, "Quantum Physics in Neuroscience and Psychology: A Neurophysical Model of Mind-Brain Interaction," Philosophical Transactions of the Royal Society of London B, Vol. 360, No. 1458, 2005, pp. 1309-1327.

17. R. Penrose, "The Emperor's New Mind: Concerning Computers, Minds and The Laws of Physics," Oxford University Press, Oxford, 1989.

18. S. Hameroff and R. Penrose, "Orchestrated Objective Reduction of Quantum Coherence N Brain Microtubules: The 'Orch OR' Model Forconsciousness," Mathematics and Computer Simulation, Vol. 40, No. 3-4, 1996, pp. 453-480.

19. R. Hari and O. V. Lounasmaa, "Neuromagnetism: Tracking the Dynamics of the Brain," Physics World, Vol. 13, 2000, pp. 33-38.

20. W. Tirsch, "Biomedizinische Relevanz der quantitativen EEG Analyse," LMU München, München, 2009.

21. K.-H. Fichtner, "Time Series Related to Quantum Measurements and the Quantum Zenon Effect," Jenaer Schriften zur Mathematik und Informatik, Math/Inf/02/ 2012, Faculty of Mathematics and Informatics, FSU Jena, Jena, 2012, 15 p.

22. M. Asano, M. Ohya, Y. Tanaka, A. Khrennikov and I. Basieva, "Quantum Like Representation of Baysian Updating," American Institute of Physics, Vol. 1327, No. 1, 2011, pp. 57-62.

23. M. Asano, I. Basieva, A. Khrennikov, M. Ohya and I. Yamato, "A General Quantum Information Model for the Contextual Dependent Systems Breaking the Classical Probability Law," ArXiv:1105.4769v1(quant-ph), 21 May 2011.

3
CHAPTER

TRAPPED-ION QUANTUM SIMULATION OF TUNABLE-RANGE HEISENBERG CHAINS

Tobias Graß[1]* and Maciej Lewenstein[1,2]

[1]ICFO-Institut de Ciències Fotòniques, Av. Carl Friedrich Gauss, 3, 08860 Castelldefels, Barcelona, Spain.

[2]ICREA-Institució Catalana de Recerca i Estudis Avançats, Barcelona, 08010, Spain.

ABSTRACT

Quantum-optical techniques allow for generating controllable spin-spin interactions between ions, making trapped ions an ideal quantum simulator of Heisenberg chains. A single parameter, the detuning of the Raman coupling, allows to switch between ferromagnetic and antiferromagnetic chains, and to modify the range of the interactions. On the antiferromagnetic side, the system can be tuned from an extreme long-range limit, in which any pair of ions interacts with almost equal strength, to interactions with a $1/r3$ decay. By exact diagonalization, we study how a system of up to 20 ions behaves upon tuning the interactions. We find that it undergoes a transition from a dimerized state with extremely short-ranged correlations towards a state with quasi long-range order, that is, algebraically decaying correlations. The dynamical evolution of the system after a local quench is shown to strongly vary in the two regimes: While in the dimerized limit, the excitation remains localized for long times, propagating spinons characterize the dynamics of the quasi-long-range ordered system. Taking a look onto the ferromagnetic side of the system, we demonstrate the feasibility of witnessing non-locality of quantum correlations by measuring two-particle correlators.

Keywords: quantum simulations with trapped ions spin models

1 INTRODUCTION

A paradigm system of quantum mechanics which may exhibit intriguing quantum properties like entanglement and non-locality are two spins. By increasing the number of spins, more complex behavior may emerge. In fact, a large variety of condensed matter phenomena, ranging from metal-insulator transition to superfluidity or superconductivity, are successfully described by mapping the relevant low-energy Hilbert space onto a spin model [1]. Moreover, spin models may describe spin-liquid phases which exhibit topological order. In recent years, technological progress in manipulating atoms on the quantum level has allowed to explicitly engineer spin models [2]. This has opened the opportunity for testing the foundations of quantum mechanics, for simulating complex many-body behavior, and for studying the dynamics of interacting spin chains [3–5].

A very promising quantum simulator are trapped ions, with widely spread applications ranging from relativistic quantum mechanics [6, 7], topological systems [8], fermionic lattice models [9], to neural networks [10]. Trapped ions can be prepared in such a way that their dynamics is mainly restricted to some internal states of the ions, while the external motion is cooled down to only a few phonons. The internal states then represent a (pseudo)spin, and phonon-mediated spin-spin interactions can be implemented [11, 12]. This has already led to the experimental realization of SU(2) Ising models in one and two spatial dimensions [13–16], and, very recently, to the experimental study of entanglement dynamics in Ising and XY chains [17, 18]. The implementation of more complicated spin models has been suggested, e.g. of Heisenberg and XY models [12, 19], or models with higher spin [20]. Furthermore, the tunability of the phonon-mediated interaction allows to study models with long-range interactions. Roughly, interactions with $1/r\alpha$ decay have been engineered for $0 \lesssim \alpha \lesssim 3$ [16].

This flexibility suggests an implementation of tunable-range spin models in trapped ions. While both theoretical and experimental literature so far has focussed on Ising- or XY-type quantum simulations [13–19, 21], here we consider a trapped-ion implementation of the Heisenberg model. In

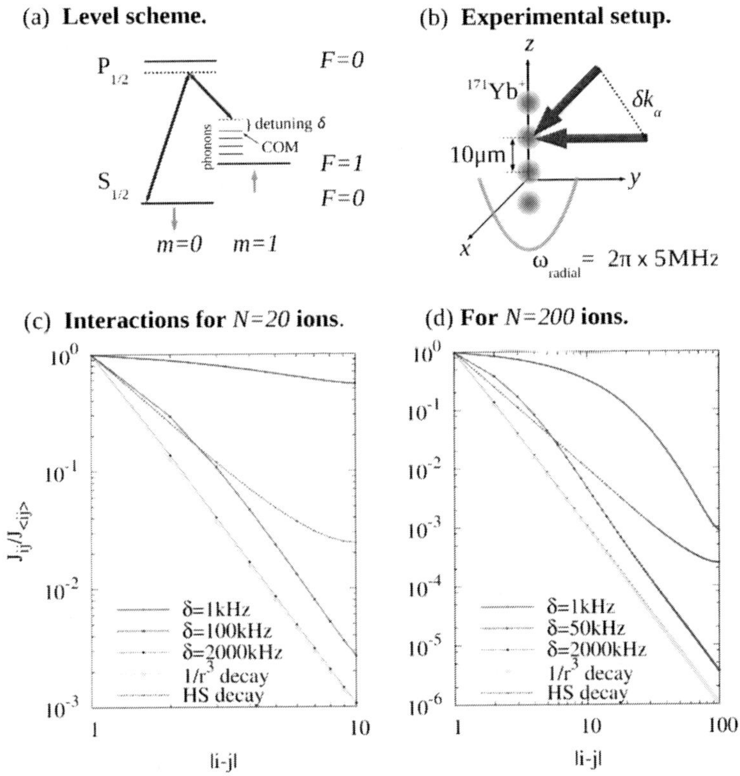

Figure 1 . ENGINEERING SPIN-SPIN INTERACTIONS. (A), (B) Level scheme and setup for a possible implementation of spin-spin interactions in $^{171}Yb^{+}$. **(C), (D)** Interaction strengths *Jij* between one ion in the center and the other ions, for (c) *N*=20 or (d) *N*=200, and different detunings from the center-of-mass (COM) mode. We have used the parameters specified in (b). Interactions are compared with the interactions of the Haldane-Shastry (HS) model (brown lines), and with $1/r3$ interactions (green lines).

particular, by assuming an experimental setup as sketched in Figure 1(a), (b), we study the influence of a single control parameter on the quantum simulation, the detuning of the Raman coupling. As shown in Figure 1(c), (d), this parameter controls the range of the interactions. Modifying it may bring our quantum simulation close to different antiferromagnetic variants of the Heisenberg model: the Haldane-Shastry model [22–24], the Majumdar-Ghosh model [25], or the Lipkin-Meshkov-Glick model [26]. While the Haldane-Shastry model describes antiferromagnetic order with algebraic decay of correlations, the Majumdar-Ghosh model provides a parent Hamiltonian for a fully dimerized ground state, that is a ground state with extremely short-range correlations. In the Lipkin-Meshkov-Glick model, spin-spin interactions are independent from distance, and the system is thus described in terms of global spin operators.

In this paper, we first provide an overview of the different models in Section 2. We then discuss the ionic setup in Section 3. In Section 4, we show that by increasing the range of interactions, a transition from a Haldane-Shastry-like quasi-long-range order to a Majumdar-Ghosh-like dimerized order can be observed in the ionic system. In Section 5, we apply different quenches to the system, and study the subsequenct dynamical evolution of the system. We consider local quenches, e.g. by a sudden application of a magnetic field to a single ion, and show that the dynamics in the dimerized regime differs significantly from the one in a system with quasi-long-range order. While in the former case the excitations remain localized for a long time, in the latter they spread in good agreement with the velocity obtained analytically for the Haldane-Shastry model. For global quenches, realized by apruptly changing the range of interactions and thereby sweeping through the dimerization transition, we study the correlation spreading, which is found at an increased velocity. In Section 6, we consider the Lipkin-Meshkov-Glick limit on the ferromagnetic side, which has been suggested for witnessing non-locality of quantum correlations [27]. Finally, we provide a summary in Section 7. Moreover, our paper contains two appendices: In Appendix 1, we provide details to the ionic setup, and in Appendix 2, we discuss the relation between the ionic model and a spin model.

2 THE MODELS

The Hamiltonian of a Heisenberg chain of N spin-1/2 particles generally reads

$$H = \sum_{i,j}^{N} \sum_{\alpha=x,y,z} J^{(i,j)} S_\alpha^{(i)} S_\alpha^{(j)}, \tag{1}$$

where $S_\alpha^{(i)} = \frac{\hbar}{2}\sigma_\alpha^{(i)}$, and $\sigma_\alpha^{(i)}$ denotes a Pauli matrix for the spin at position i. In this paper we focus on the antiferromagnetic side of this model, that is, the model with interaction strengths $J^{(i,j)}$ >0.

The functional behavior of $J^{(i,j)}$ as $|i-j|$ may crucially influence the physics of the model. Let us in the following discuss the different cases which are important for our application.

2.1 Haldane-Shastry model

The Haldane-Shastry model is paradigmatic for Heisenberg chains with antiferromagnetic long-range interactions. It stands out through its analytical solution in terms of a Gutzwiller ansatz [22, 23]. The model is characterized by interactions which decay quadratically with the distance *dij* between the two spins, that is, *J*(*i*,*j*)*α*≡*Jd*−2*ij*. Periodic boundary conditions are imposed by arranging the spins on the unit circle, that is, with positions $z_k = \exp[\frac{2\pi i}{N}k]$ In this case, $d_{ij} = 2\sin(\frac{\pi}{N}|i-j|)$. For a chain with even number of spins, the ground state reads

$$|\Psi_0\rangle = \sum_{\{z_1,\ldots,z_M\}} \Psi_0(z_1,\ldots,z_M) S^+_{z_1}\cdots S^+_{z_M}|\downarrow\downarrow\cdots\downarrow\rangle, \tag{2}$$

with $M = N/2, S^+_{z_i} = |\uparrow\rangle\langle\downarrow|_{z_i}$ a raising operator of the spin at position *zi*, and the coefficients of each Fock state given by the wave function

$$\Psi_0(z_1,\ldots,z_M) = \prod_{i<j}^{M}(z_i - z_j)^2 \prod_{i=1}^{M} z_i. \tag{3}$$

This function has a remarkable similarity to the Laughlin wave function for two-dimensional systems in the fractional quantum Hall regime [28]. The similarities can be extended to the excited states of the model which are spinons with certain anyonic properties: Obtained as a superposition of spin flips, excitations live in an integer-spin Hilbert space, but, only occuring pairwise, each spinon carries half-integer spin. In that sense, the spinon represents a quasiparticle with a fractional quantum number. Similarly to excitations in Laughlin's state, the spinon wavefunctions can be obtained from the ground state by formally piercing a hole localized to position η into it:

$$\Psi_\eta(z_1,\ldots,z_M) = \prod_i(\eta - z_i)\Psi_0(z_1,\ldots,z_M). \tag{4}$$

Assuming now an odd number of spins, the state of a $\downarrow$-spinon localized at η is given by $|\Psi_{\eta\downarrow}\rangle = \sum_{\{z_1,\ldots,z_M\}} \Psi_\eta(z_1,\ldots,z_M) S^+_{z_1}\cdots S^+_{z_M}|\downarrow\downarrow\cdots\downarrow\rangle$, where

M=(N−1)/2. By Fourier transformation of the localized spinon wave function, one obtains the wave function for a propagating spinon, which is an the exact solution of the model:

$$\Psi_{p_m}(z_1,\ldots,z_M) = \sum_{\eta}(\eta^*)^m \Psi_\eta(z_1,\ldots,z_M), \tag{5}$$

where m is an integer defining the momentum $p_m = -(2\pi/N)m + \pi(1 - 1/2N)$. Evaluating the energies of Ψ0 and Ψpm, one obtains the spinon excitation energy as [24]:

$$\epsilon(p_m) = \frac{JN^2}{4\pi^2}\left[\frac{p_m}{2}(\pi - p_m) + \frac{\pi^2}{8N^2}\right]. \tag{6}$$

From this, we read off the spinon velocity $v_s = \frac{JN^2}{8\pi}$

. We will later compare this result with the spinon velocity in an open chain. Therefore we have to note that the distance of neighboring spins on the unit circle is $2\pi/N$, while on the open chain we conveniently take it as a unit of length. Accordingly, one has to rescale $J \to J\frac{4\pi^2}{N^2}$, and the spinon velocity reads

$$v_s = J\frac{\pi}{2}. \tag{7}$$

Returning to the ground state properties of the Haldane-Shastry model, we shall consider its spin correlations. It has been shown analytically that the model supports correlations with a power-law decay [29]

$$\left\langle \sigma_z^{(i)}\sigma_z^{(j)}\right\rangle \propto \frac{(-1)^{|i-j|}}{|i-j|}. \tag{8}$$

With this criterion, the Haldane-Shastry model, despite the long-range interactions, supports the same quantum phase as the Heisenberg chain with nearest-neighbor interactions, also characterized by quasi long-range spin order. In the next subsection, we will introduce a model which is in contrast to this behavior.

2.2 Majumdar-Ghosh model/J1–J2 model

The Heisenberg model with nearest-neighbor and next-nearest-neighbor interactions, J1 and J2, is called J1–J2-model. It is known to exhibit a dimerization transition when $J_2/J_1 \gtrsim 1/4$ [30–32]. The physical consequences of the dimerization become clearest for J2=0.5J1 and periodic boundary conditions. Then the model becomes identical to the Majumdar-Ghosh model [25], which is solved by two degenerate ground states $|\Psi+\rangle$ and $|\Psi-\rangle$. These states are obtained by bringing every second nearest-neighbor pair into a spin singlet configuration, such that the total state is a product over singlet bonds:

$$|\Psi_{\pm}\rangle = \prod_{i=1}^{N/2} \left(|\uparrow\downarrow\rangle_{2i,2i\pm1} - |\downarrow\uparrow\rangle_{2i,2i\pm1}\right)/\sqrt{2}. \tag{9}$$

It is obvious that in such dimer state, any spin is fully correlated to one nearest neighbor, but fully uncorrelated with the other spins. In other words, no long-range spin correlations exist. Note that for a system with open boundary conditions, the spin at position 1 and the spin at position N do not interact, and therefore the ground state is uniquely given by $|\Psi-\rangle$

.

2.3 Lipkin-Meshkov-Glick model

By strengthening interactions between distant spins, one approaches the Lipkin- Meshkov-Glick model [26] in which interactions are spatially independent, that is, $J^{(i,j)}$. With this, the Hamiltonian is rewritten as $H = J(\mathbf{S}^2 - \frac{3}{4}N)$ with the total spin operator $\sum_i(\sigma_x^{(i)}, \sigma_y^{(i)}, \sigma_z^{(i)})$ /For antiferromagnetic interactions, any singlet state is thus a ground state, leading to a huge degeneracy. The number of singlet states formed by N spin-1/2 particles is given by the Catalan number

$$C_n = \frac{N!}{(\frac{N}{2})!(\frac{N}{2}+1)!}. \tag{10}$$

However, this degeneracy is lifted by any small spatial dependence of the interactions.

As we will see below, the ionic systems approaches the Lipkin-Meshkov-Glick limit when the Raman coupling is close to resonance with the center-of-mass phononic mode. By going through the resonance, one is able to swap the sign of the interactions, thus both the antiferromagnetic and the ferromagnetic Lipkin-Meshkov-Glick model can be realized. On the ferromagnetic side, the ground states are the symmetric Dicke states, that is, symmetric superpositions of all states with a fixed spin polarization, that is with a fixed number $N\uparrow(N\downarrow)$ of ↑-(↓-)spins. The unnormalized, symmetric Dicke states read

$$|D_{N_\uparrow,N_\downarrow}\rangle \equiv \sum_{\{i_1,\ldots,i_N\}} |\uparrow\rangle_{i_1}\cdots|\uparrow\rangle_{i_{N_\uparrow}}|\downarrow\rangle_{i_{N_\uparrow+1}}\cdots|\downarrow\rangle_{i_{N_\uparrow+N_\downarrow}}. \tag{11}$$

Apparently, the ground state degeneracy is $(N+1)$

-fold, and, as each Dicke state has a different spin polarization, it can be lifted by a polarizing field term $\sim h\sum_i \sigma_z^{(i)}$. While such term will make the fully polarized state |□↓···□↓⟩ the unique ground state, one can also obtain the polarized Dicke state $|DN/2,N/2\rangle$ as the unique ground state by reducing the Heisenberg XXX interactions to XX interactions. The Hamiltonian then reads

$$H_{\mathrm{LMG}} = J\sum_{i<j}\left(S_x^{(i)}S_x^{(j)} + S_y^{(i)}S_y^{(j)}\right) + h\sum_i S_z^{(i)}, \tag{12}$$

with $|DN/2,N/2\rangle$ the unique ground state for $J<0$ and $h=0$.

3 THE IONIC SYSTEM

Tunable-range Ising models have already been implemnted in trapped ions [13, 14]. In these experiments, two-levels ions are confined to a line by a Paul trap, and their motional state is cooled to only a few phonons. Strong spin-spin interactions are then achieved by applying state-dependent forces on the ions. In Ref. [13], Ising-type interactions of the form $\sigma_z^{(i)}\sigma_z^{(j)}$ are generated, whereas Ref. [14] describes the production of interactions of the form $\sigma_x^{(i)}\sigma_x^{(j)}$ x or $\sigma_y^{(i)}\sigma_y^{(j)}$. As reviewed in Ref. [33], both approaches have basically the same footing, and it is possible to combine them. Instead of an Ising coupling, one then obtains, in the first place, an *XYZ*-model. By making all interactions equal, one gets the Heisenberg model.

An important difference between the ionic system and the ideal models discussed above are the boundary conditions: For the most feasible experimental implementation, they are open, while periodic boundary conditions are convenient for a theoretical description. In general, the effect of boundary conditions is minimized by scaling up the system. For systems of up to N=20ions, we will in the following discuss how close the connection to the Haldane-Shastry model and the Majumdar-Ghosh model can be made by tuning the range of the interaction.

To this goal, let us first briefly review how the desired spin-spin interactions can be engineered. For each coupling, $\sum_{i<j}\sigma_\alpha^{(i)}\sigma_\alpha^{(j)}$, a pair of Raman lases is set up, as depicted in Figure 1(a), (b). Making several assumptions, which are sketched in the Appendix 1 and detailed in Ref. [33], the Hamiltonian for the interaction of the ions with the photons is given by $H(t)=\sum_{\alpha=\mathrm{x,y,z}}h_\alpha(t)$, with

$$h_\alpha(t)=\frac{\hbar\Omega_\alpha}{2}\sum_{i=1}^{N}\sum_{m}\eta_m\left(\hat{a}_m e^{-i(\omega_\alpha-\omega_0-\omega_m)t}+\mathrm{H.c.}\right)\sigma_\alpha^{(i)}, \tag{13}$$

for α=x,y, and

$$h_z(t)=\frac{i\hbar\Omega_z}{2}\sum_{i=1}^{N}\sum_{m}\eta_m\left(\hat{a}_m e^{-i(\omega_z-\omega_m)t}+\mathrm{H.c.}\right)\sigma_z^{(i)}. \tag{14}$$

Here, ħω0 is the energy difference between the two internal levels, aˆm is the annihilation operator for phonons denoted by m and with frequency ωm. The Rabi frequencies of the couplings are denoted by Ωα, and $\omega\alpha$ are the frequencies of the fields. Furthermore, the strength of each coupling depends on the Lamb-Dicke paremeters ηm, which are explicitly defined in the Appendix 1. For all couplings α, the wave vector difference of the photons, $\delta\mathbf{k}\alpha$, is assumed to be transverse to the ion chain, so the sum over m reduces to a sum over N transverse modes.It has been shown in Ref. [14], for a system with a single coupling term $h\alpha$, that the time evolution can be made identical to the one of a spin system with the Hamiltonian

$$H_\alpha=\frac{1}{\hbar}\sum_{i\le j}J_\alpha^{(i,j)}S_\alpha^{(i)}S_\alpha^{(j)}, \tag{15}$$

where the spin-spin interaction strength is given by

$$J_\alpha^{(i,j)}=\Omega_\alpha^2\sum_m\frac{\eta_{m\alpha}^{(i)}\eta_{m\alpha}^{(j)}}{\tilde{\omega}_\alpha-\omega_m}. \tag{16}$$

Here, $\tilde{\omega}_\alpha \equiv \omega_\alpha - \omega_0$ for α=x,y, whereas $\tilde{\omega}_\alpha \equiv \omega_\alpha$ for α=z. Some details of the derivation of this formula are provided in the Appendix 2. In particular, we generalize to the case of more than one coupling, and show that the couplings can be chosen such that they do not interfere. The Hamiltonian then is effectively given by

$$H = \frac{1}{\hbar} \sum_{\alpha} \sum_{i \leq j} J_\alpha^{(i,j)} S_\alpha^{(i)} S_\alpha^{(j)}, \tag{17}$$

with all $J_\alpha^{(i,j)}$ given by Eq. (16). These interactions can be tuned by varying the frequency of the Raman laser. Choosing it close to the frequency of the center-of-mass mode, the strength of the induced interactions will barely depend on the particles' position, $J^{(i,j)} \approx J$. The sign of J can be made positive (negative) by tuning above (below) the center-of-mass frequency. Close to resonance, the system is similar to the Lipkin-Meshkov-Glick model. Due to the strong nearest-neighbor and next-nearest-neighbor interactions this limit is somewhat similar to the dimerized J_1-J_2-model.By increasing the detuning from the center-of-mass mode, $J^{(i,j)}$is made spatially dependent, and in the limit of large detuning, a $1/r^3$ decay can be achieved. For intermediate values of the detuning, the interactions may approximate a quadratic decay for sufficiently small r, as shown in Figure 1(c), (d). The most dominant interactions then agree quantitatively well with the interactions of the Haldane-Shastry model. Further techniques could be applied in order to achieve a better agreement of the ion setup with a particular model, e.g. by individually addressing of the ions [34]. However, here we restrict ourselves to the simplest implementation which already exhibits rich physics.

The feasibility of the experimental implementation depends on different factors which potentially reduce the fidelity. In the first place, one should note that the effective Hamiltonian (17) has been obtained by neglecting a residual spin-phonon coupling. This implies an error $\epsilon_{sp} \approx 4(1+\bar{n})\eta^2$, with $\bar{n}$ the mean phonon number [12]. This error is small in the Lamb-Dicke regime, $\eta \ll 1$, typically of the order of few percents [15]. Another source of error, which becomes particularly important for small detunings, are Raman-induced phonon excitations, with a probability given by $p_{ph}=(\eta\Omega/\delta)^2 \sim J/\delta$. Keeping this error small at a given detuning therefore sets an upper limit to the coupling parameter J. On the other hand, J should be sufficiently large compared to imperfections like field imhomogenities or spontanous emission rates. The recent experiments in Maryland [17] and in Innsbruck [18] with up to 15 trapped ions operate at a $J_{max} \sim 400$Hz, much larger than the estimated inhomogenities <20Hz [18]. We will find in the next section that dimerization occurs at a detuning of the order of 10kHz, that is, slightly below the lowest detuning realized in Ref. [18]. In this regime, photon errors are estimated to be $p$ph=0.04.

An important part of trapped-ion quantum simulations are the well developped read-out techniques. In particular, it is possible with single-site precision to excite ions from one spin level to some excited state, and detect the subsequent fluorescence. This allows local $\sigma_z^{(i)}$ measurements. Repeated measurements at different i will allow to evaluate correlation functions, e.g. $\langle \sigma_z^{(i)} \sigma_z^{(j)} \rangle$, as has been demonstrated, even with temporal resolution, in Refs [17, 18].

A key issue is the scalability of spin model quantum simulations. This topic has been discussed in Refs [15, 35], finding only a slow $N^{1/3}$ growth of the errors due to spontaneous emission. Nevertheless, clearly outperforming classical computing resources with trapped ions may require new architectures, as the quantum simulation of a two-dimensional system with hundreds of ions in a Penning trap [16].

4 DIMERIZATION TRANSITION

We have studied by means of exact diagonalization chains of up to 20 ions. The only tunable parameter in our study is the detuning $\delta \equiv \omega\tilde{}\alpha - \omega$COM , where ωCOM refers to the frequency of the center-of-mass mode, that is, the transverse mode of largest energy. For convenience, we have chosen Yb+171 ions, with equilibirum distance of 10μ m at a radial trap frequency ωtrap=2π×5MHz. As shown in Figure 1(c), (d), for very small detuning, δ=1kHz, the system is close to the Lipkin-Meshkov-Glick limit: Any pair of ions interacts with almost the same strength. For large detuning, $\delta \gtrsim$1,000kHz, the interactions decay with 1/r3 . For intermediate values, the interactions between near neighbors becomes similar to the Haldane-Shastry interactions.

In Figure 2(a), we plot the overlap of the ground state of 16 ions with the ground state manifold of the Majumdar-Ghosh model, and the Haldane-Shastry model as a function of the detuning. In the limit of small detuning, the ground state is almost fully dimerized. Increasing the detuning, the ground state of the Haldane-Shastry model becomes more relevant. One has to note, however, that (for 16 particles) this state is far from being orthogonal to the dimerized subspace. In fact, it has itself an overlap of 0.73 with the dimerized manifold.

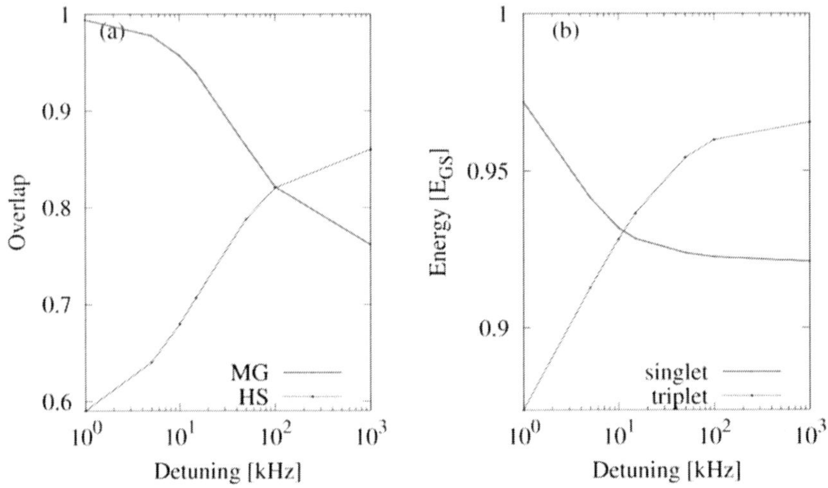

Figure 2. SYSTEM BEHAVIOR UPON TUNING. (A) Overlap of the ground state of N=16

ions with the ground state manifold of the Majumdar-Ghosh model, and the Haldane-Shastry model as a function of the detuning. **(B)** Energies of the first singlet and the first triplet excitation (normalized by the ground state energy), as a function of the detuning.

A sharp criterion for the transition from a dimerized state to a long-range ordered state can be inferred from Figure 2(b), where the energy of the lowest singlet and the lowest triplet excitation is plotted. For the $J1-J2$

model, it is known that the crossing of these energies mark with high precision, even in small systems, the dimerization transition [36]. In our case, the triplet becomes the low-lying excitation for δ>10kHz. In this context, we also note that, at δ=1kHz, in total 118 singlet states have lower energy than the triplet state. Still, this is far from $C16$=1,430, the number of singlet states providing the ground state manifold of the antiferromagnetic Lipkin-Meshkov-Glick model. But in contrast, at δ=5kHz , the lowest triplet state already reaches the 7th position in the energy spectrum.

4.1 Entanglement entropy

A dimer is a pair of maximally entangled spins. A measure which localizes the entanglement within a system, and which is thus able to identify dimerization, is the entanglement entropy. This quantity considers bipartitions of the system, and measures the entanglement between the two subsystems. In our case the spins form an one-dimensional array, and it is thus natural to consider the bipartitions $A=\{1,\ldots,\ell\}$ and $B=\{\ell+1,\ldots,N\}$. The entanglement entropy is then a function of ℓ, defined as

$$\mathcal{S}(\ell) = -\operatorname{Tr}\left[\hat{\rho}(\ell)\log_2 \hat{\rho}(\ell)\right], \quad (18)$$

where $\rho\hat{}(\ell)$ is the density matrix of the subsystem A.

The nearest-neighbor dimerized states, $|\Psi\pm\rangle$ of Eq. (9), are characterized by a strongly alternating behavior of the entanglement entropy, as shown in Figure 3: A cut through every second bond will yield strong entanglement, $S(\ell)=1$, as the spins ℓ and $\ell+1$ are dimerized, whereas on the other bonds no quantum information is shared, $S(\ell)=0$. Such alternations, however, are not present for the Majumdar-Ghosh dimer with periodic boundary conditions [37]. In that case, both dimer states $|\Psi+\rangle$ and $|\Psi-\rangle$, will equally contribute to the ground state manifold, and thus completely wash out the pattern. It has been discussed in Ref. [38] for a broader class of spin models that alternating behavior is characteristic for systems with open boundary conditions, though typically with a much smaller amplitude than in the case of the Majumdar-Ghosh chain.

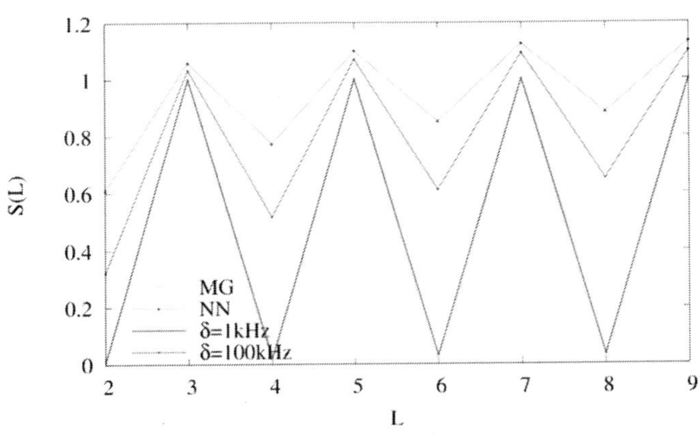

Figure 3. ENTANGLEMENT ENTROPY. We plot the entanglement entropy of the ionic system with N=18 for detuning δ=1kHz and δ=100kHz , and for ideal Heisenberg chains with open boundary conditions and nearest-neighbor interactions (NN) and Majumdar-Ghosh (MG) interactions.

In Figure 3, we plot the entanglement entropy of different systems with N=18 spins: the ionic system with detuning δ=1kHz and δ=100kHz, an ideal Heisenberg chain with nearest neighbor interactions, and the Majumdar-Ghosh chain. In all cases, we have applied open boundary conditions, and accordingly we find alternating behavior of the entanglement entropy. As expected, these alternations are strongest for the Majumdar-Ghosh chain, and weakest for the nearest-neighbor Heisenberg model. The entanglement entropy of ionic systems with δ=1kHz behaves very similar to the entanglement entropy of the Majumdar-Ghosh model, proving the dimerized nature of the phase. At δ=100kHz , the curve alternates less and comes closer to the entanglement entropy of the nearest-neighbor Heisenberg model. This shows that the tendency of nearest neighbors to form singlets is still present, but also more remote spins become entangled.

4.2 Spin-spin correlations

An experimentally accessible quantity which nicely displays the different entanglement properties are spin correlations. While correlations will, in principle, depend on the position of the spins, we define an average which depends only on the distance d between the spins:

$$C(d) = \frac{1}{N-d}\sum_{i=1}^{N-d}\left\langle \sigma_z^{(i)}\sigma_z^{(i+d)}\right\rangle. \tag{19}$$

To some extent this restores periodic boundary conditions. Furthermore, we perform a finite-size scaling (taking into account all even system sizes from N=10 to N=20).

The results, for δ=100kHz , are shown in Figure 4. The sign of the correlations alternates with odd/even d. The correlations compare well with the correlations of the Haldane-Shastry model, despite the different boundary conditions. The decay is slightly too fast, but through finite-size scaling a slightly better agreement is obtained. One should also note that for large d, comparable to the system size, edge effects are not averaged out by the definition of Eq. (19). The plot in Figure 4(b) demonstrates that the decay can be modeled by a power-law: $C(d)\propto(-1)dr-\alpha$, with α=1.15±0.05.

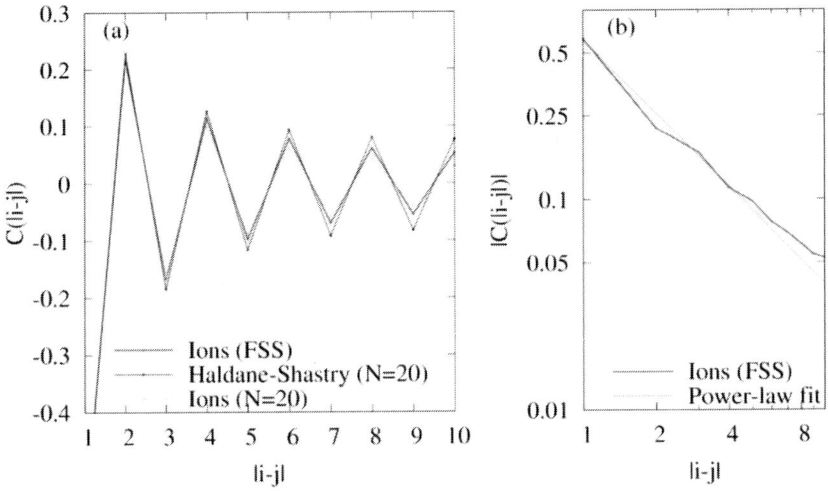

Figure 4. CORRELATIONS FOR LARGE DETUNING. (A) Correlations of the ionic system (δ=100kHz) after finite-size scaling (FSS), and for N=20, in comparison with the correlations in the Haldane-Shastry model (with periodic boundary). **(B)** Power-law fit to the correlations shows a 1/ $r\alpha$ decay, with α=1.15±0.05.

On the dimerized side, we find correlations as shown in Figure 5. Since $C(d)$ as defined in Eq. (19) would average out the large dimer correlations with the essentially uncorrelated bonds between two dimers, we have defined

$$C^{\mathrm{odd}}(d) = \frac{1}{i_{\max}} \sum_{i=1}^{i_{\max}} \left\langle \sigma_z^{(2i-1)} \sigma_z^{(2i-1+d)} \right\rangle, \tag{20}$$

where $i$max=(N−d)/2 for d even, and $i_{\max}$=(N−d+1)/2 for d odd. With such definition, we find that Codd(d) takes its maximum value −1/4 for d=1, demonstrating the strong anticorrelations between every second nearest-neighbor pair. For larger distances, the value rapidly decreases. The sign alternates for odd/even d. As shown in Figure 5, the decay takes place exponentially.

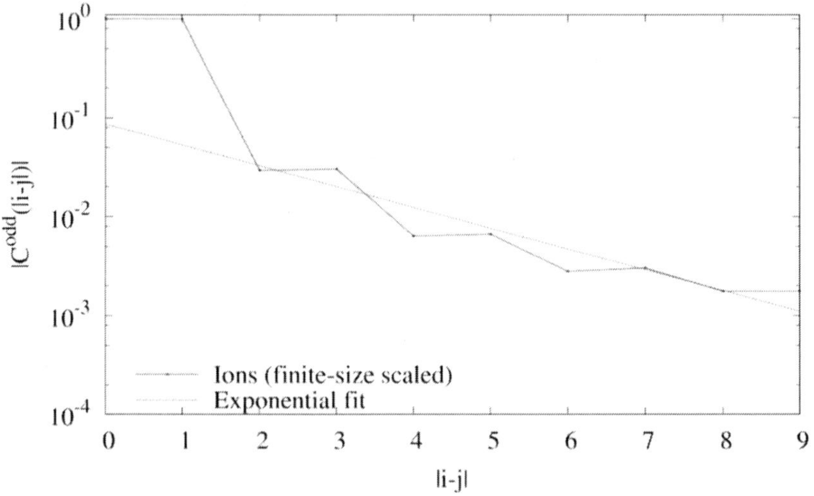

Figure 5. CORRELATIONS FOR SMALL DETUNING. Staggered correlations (defined in Eq. (20) of the ionic system (δ=1kHz) after finite-size scaling (FSS). An exponential fit models well the decay.

5 QUENCH DYNAMICS

So far we have considered ground state properties of the trapped ion quantum simulation of the Heisenberg model. In two very recent experiments [17, 18], trapped ions have been demonstrated to be particularly useful for studying also the dynamics of interacting spins. This can be done by exciting the system through either a local or a global quench. In the following, we will investigate the system's response to local quenches in both the dimerized and the quasi-long-range ordered regime. Afterwards, we study a global quench through the dimerization transition.

5.1 Local quenches

We consider a system which initially is prepared in the ground state of *H* given by Eq. (17), that is, a system with pure spin-spin interactions. As discussed in the previous section, for sufficiently small detuning, this is a dimerized state, whereas larger detuning yields a ground state with algebraically decaying correlations. In both cases, the spin polarization $\langle\sigma_z^{(i)}\rangle$is zero on all sites *i*. We then create locally a finite spin polarization by abruptly switching on a magnetic field term at one site, for instance at the position of the outermost left spin, *i*=1. Subsequently, we evolve the system under the new Hamiltonian $H' = H + BS_z^{(1)}$. The evolution of the spin polarization is shown for a system of *N*=12 spins in Figure 6.

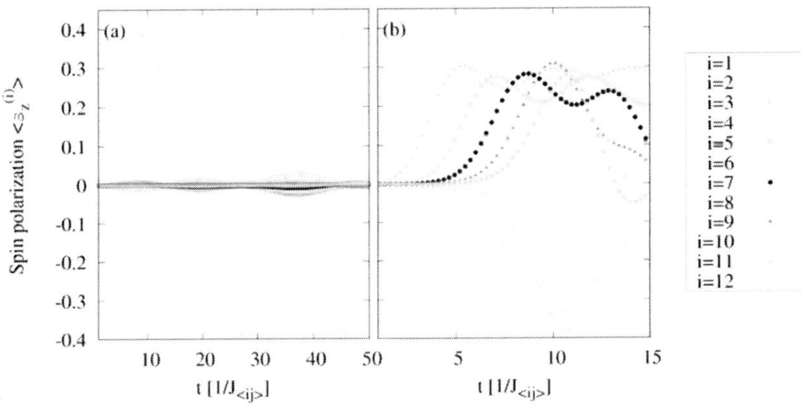

Figure 6 . SPIN POLARIZATION AFTER A LOCAL QUENCH: At time *t*=0 a "magnetic field" along the *z*-axis is switched on abruptly at site *i*=1. The subsequent dynamics of the spin polarization is shown for spin-spin interactions in **(A)** the dimerized limit (detuning 1kHz), that is extremely long-range, and **(B)** the 1/*r*3 limit (detuning 10MHz). Counterintuitively, for extremely long-range interactions the information remains localized for a long time, except for the immediate communication with the dimerized neighbor spin. In (b), we find spreading of the information in form of two spinons.

In the dimerized system, Figure 6(a), the magnetic field induces a spin oscillation of the first spin. Simultaneously, also the second spin performs an oscillation in the opposite direction. This behavior is an apparent consequence of the strong anticorrelations between the two spins. Interestingly, all other spins remain almost unaffected by the magnetic field for a long time.

A very different behavior is found in the quasi long-range ordered system. Surprisingly, despite the shorter range of interactions, the local perturbation now influences distant spins in significantly less time. Again, the quench on site *i*=1 immediately affects the spin at site *i*=2, as a consequence of strong anticorrelations between neighboring spins.

However, as seen in Figure 6(b), the spin polarization also propagates quickly to further spins, leading to positive (negative) spin polarization of the odd (even) spins. Therefore, we may interpret the excitation as a pair of spinons, that is, two spin flips on neighboring sites traveling through the chain. In Figure 7, we further analyze our data by determining the time at which a spin is "activated". Herefore, we demand that $|\langle\sigma_z^{(i)}\rangle| > 0.02$. .With this, we are able to define a spinon velocity, plotted in Figure 7(b). On average, this velocity compares well with spinon velocity in the Haldane-Shastry model, $v_s = J\frac{\pi}{2}$. At any point in time, the velocities remain below the Lieb-Robinson bound for *Sz* in the nearest-neighbor Heisenberg model, *v*LR=*eJ*[39].

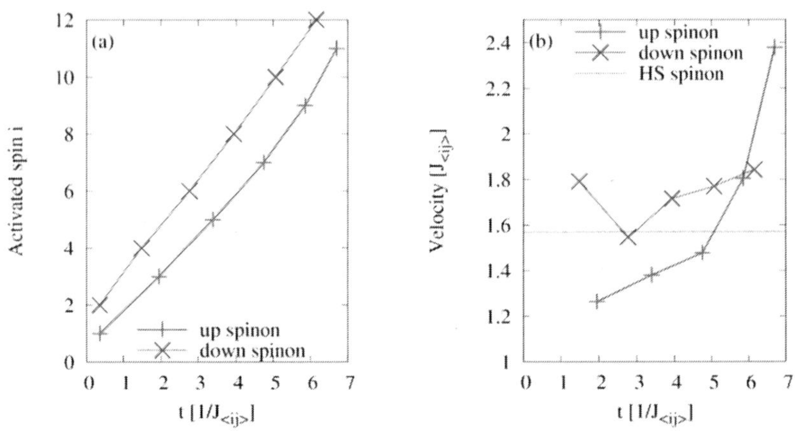

Figure 7. VELOCITY OF EXCITATIONS. (A) Position vs. time of a spin when it starts being polarized ($\langle\sigma(i)z\rangle$>0.02) in a system with 1/*r*3 decaying interaction (detuning 10MHz). **(B)** Corresponding spinon velocity as a function of time. The black line corresponds to the spinon velocity in the Haldane-Shastry model.

An alternative but similar way of performing a local quench is by a projective measurement of one spin. The subsequent dynamics is then described purely by the Hamiltonian *H*, and is found to be qualitatively similar to the dynamics described above. The main difference is that, in the system with quasi long-range correlations, the projective measurement will immediately affect all spins to a certain extent, obscuring the signal of the spinons.

While in practice it might be difficult to maintain coherence during the whole evolution shown in Figure 6(a), up to times of the order $50J^{-1}$, the striking differences between Figure 6(a) and (b) become apparent already on a much smaller time scale, say $5J^{-1}$. For typical values of J, this is of the order of 10ms, which is smaller than the duration of measurements in Ref. [18].

5.2 Global quenches

We have also studied the systems' dynamics following a global quench. Here an interesting possibility is to abruptly change the range of interactions by changing the detuning. Thereby one can quench through the dimerization transition. Starting from the dimerized ground state, where each spin is correlated only to one neighbor, one can observe the creation of entanglement between more distant spins.

As a first observation we note that the initially dimerized pairs remain strongly anticorrelated for all times. For the correlations between the first and the second spin, we have $C12(t)<-0.8\ \forall t$. In Figure 8, we show the growth of correlations between the first spin and the remaining spins, to which no correlations exist in the beginning. In analogy to the behavior of spin polarization discussed before, the spins at odd positions will, in the first place, become correlated with the first spin, while the even spins are getting anticorrelated. As a guide to the eye, the vertical lines in Figure 8 mark the "trajectory" of a signal traveling with the Haldane-Shastry spinon velocity, $vs=\pi/2$. The first line corresponds to a traveled distance of one unit of length. For subsequent lines, the distance is increased by two. As the first and the second ions are dimerized, we may always consider the distance between an ion i to the second ion, i=2. We then find that the velocity with which correlations reach the spins at i=3, i=5, and i=7 coincides well with the spinon velocity. Interestingly, the anticorrelations with the even spins are transmitted more rapidly. A remarkable increase of velocity is found towards the end of the chain: Spins at i=7, i=9, and i=11 become correlated with the first spin almost simultaneously. Even more surprisingly, the correlation with the spin at i=11 starts to grow slightly earlier than the correlation with the spin at i=9. An explanation of this behavior could be the fact that in the moment of the global quench excitations are created at all positions. Each excitation will, in the first place, build up correlations between neighboring spins, but nonlocality of correlations can provide an immediate "bridge" between spins at a larger distance.

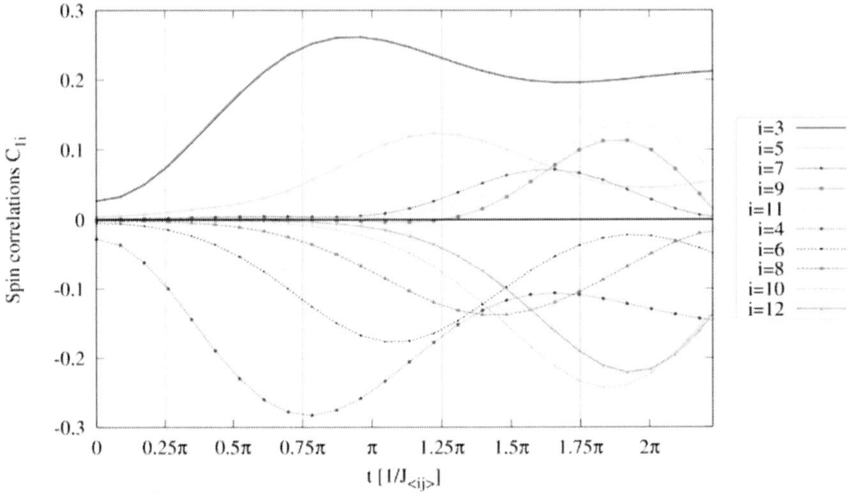

Figure 8. SPIN CORRELATIONS AFTER A GLOBAL QUENCH: At time t=0 , the system is quenched by changing the detuning from 1kHz to 10MHz, i.e., the range of interactions is changed. Before the quench, the system is in a dimerized state of localized correlations, after the quench correlations spread throughout the whole system.

6 NONLOCALITY WITNESS

In the previous sections, we have discussed one intriguing aspect of quantum mechanics which can be studied in the ionic system: entanglement. A strongly related and particular striking feature of entanglement is the nonlocality of correlations [40]. Nonlocality can be defined as the impossibility of classically simulating the outcome of a local measurement while a second, remote measurement is performed unless some information obtained during this measurement is shared. As first shown by JS Bell, nonlocality is witnessed by the violation of some inequalities which local correlations have to fulfill [41]. In general, unfortunately, Bell inequalities depend on various high-order correlation functions, and it is thus difficult to witness nonlocality in a system. In Ref. [27], however, it has been proposed to detect nonlocality by measurement of two-body correlators only. As a specific example, a Bell inequality has been given which is violated by the spin-polarized Dicke state $|DN/2,N/2\rangle$. It reads

$$\frac{N(N-1)}{4}S_{00} + \frac{N}{2}S_{10} - \frac{1}{2}S_{11} + \frac{1}{4}N(N-1)(N+2) \geq 0. \tag{21}$$

Here, $S_{ij} \equiv \sum_{k \neq l} \langle \hat{m}_i^{(k)} \hat{m}_j^{(l)} \rangle$ are the correlations of two measurements $\hat{m}$ taken at sites k and l. On each site, one may take $\hat{m}_0 \equiv \sigma_z$ and $\hat{m}_1 \equiv \cos\theta\sigma_z + \sin\theta\sigma_x$.. If a θ for which inequality (21) is violated exists, the system must have non-local correlations.

While inequality (21) turns out to be unable to detect nonlocality on the antiferromagnetic side of the ionic setup, a chance for violation exists on its ferromagnetic side. For sufficiently small detuning, the sytem is in the Lipkin-Meshkov-Glick limit of spatially independent interactions, and for $J<0$, states with maximal spin are ground state of the Hamiltonian $H \approx J(\mathbf{S}^2 - \frac{3}{4}N)$ The ground state manifold is thus given by all $N+1$ Dicke states. To lift the degeneracy, one could apply a magnetic field: A field in z-direction yields either $D_{0,N}$ or $D_{N,0}$ as the unique ground state. The state $|D_{N/2,N/2}\rangle$ becomes unique ground state if a staggered magnetic field along z-direction is applied. Alternatively, even without applying a symmetry-breaking field, the degeneracy is lifted in favor of $|D_{N/2,N/2}\rangle$ by switching off the interactions $J_z^{(i,j)} \to 0$. The system is then govered by HLMG given in Eq. (12).

In these cases, non-locality can be proven via inequality (21), as the spin-spin correlation of $|D_{N/2,N/2}\rangle$

are given by

$$S_{xx} \equiv \sum_{i \neq j} \langle \sigma_x^{(i)} \sigma_x^{(j)} \rangle = \frac{N^2}{2}, \tag{22}$$

$$S_{zz} \equiv \sum_{i \neq j} \langle \sigma_z^{(i)} \sigma_z^{(j)} \rangle = -N, \tag{23}$$

$$S_{zx} \equiv \sum_{i \neq j} \langle \sigma_z^{(i)} \sigma_x^{(j)} \rangle = 0. \tag{24}$$

With this, inequality (21) is violated. To assert the feasibility of such non-locality witness in the ionic system, we have calculated the ground state of a system of 14 ions, with $J_z^{(i,j)} = 0$, while $J_x^{(i,j)}$ and $J_y^{(i,j)}$ given by Eq. (16). Indeed, the system is found in the Dicke state $|D_{N/2,N/2}\rangle$ with fidelity >0.99 for $|\delta|<1$kHz. By increasing the absolute value of the negative detuning, we find that, while S_{zz} and S_{zx} remain constant, S_{xx} decreases. Thus, for S_{xx} below a critical value $S^{crit}{}_{xx}=N(N-2)/2$, inequality (21) will not be violated anymore. For N=14, we numerically found Sxx=89.6>$S^{crit}{}_{xx}$ at δ=−2.8kHz, while S_{xx}=73.3<$S^{crit}{}_{xx}$ at δ=−3.0kHz. In the same parameter range, the overlap of the ground state with the Dicke state drops suddenly: While it remains above 0.95 up to $|\delta|$=2.5kHz, it reaches zero for $|\delta|$=3.2kHz.

In this regime of relatively large detuning, other phonons than the center-of-mass mode strongly affect the spin-spin interactions. This gives rise *t* very peculiar interaction patterns consisting of attractively and repulsively interacting pairs. It would, however, be striking if non-locality of correlations could also be witnessed in a system where interactions are short-ranged. Interestingly, we find that the short-range ferromagnetic Heisenberg chain behaves exactly the same way as the long-range chain does: N+1Dicke states form a ground state manifold in which degeneracies can be lifted by magnetic fields, in particular, in favor of $|DN/2,N/2\rangle$ by a staggered magnetic field. Certainly, the ionic setup discussed here does not immediately allow for implementation of the ferromagnetic short-ranged Heisenberg model due to the mentioned mixing of different phonon modes, but it could be achieved by switching from transverse to longitudinal phonons as transmitter of interactions. In that case, the center-of-mass mode is lowest in energy, and the negative detuning will not interfere with other modes. Moreover, additional control could be implemented through additional Raman couplings [34]. Alternatively, also atoms in optical waveguides or phononic crystals allow for implementing spin models with controllable interactions, and could be tuned into a ferromagnetic short-range regime [42]. We also note that by using adiabatic passage techniques the Dicke state $|D_{N/2,N/2}\rangle$has recently been produced in trapped ion systems [43].

7 SUMMARY

We have simulated a quantum simulation of the Heisenberg model with trapped ions. In particular, we have investigated the influence of a single control parameter, the detuning, on the system. This parameter allows to tune the range of the interactions. We have shown that this can trigger a dimerization transition. To demonstrate dimerization, we have calculated the entanglement entropy and two-spin correlation functions in systems of up to N=20

ions. Studying the response to local quenches, we were able to distinguish between the two regimes also by the system's dynamics: For long times, local excitations do not spread in the dimerized system. In contrast to this, they are carried through the whole chain in form of spinons in the regime with quasi-long-range order. We also have calculated the spreading of correlations following a global quench, and found some signature of non-locality. Apart from this, we have explored the possibility of witnessing non-locality by measurement of two-spin correlation functions on the ferromagnetic side of the model. In summary, the trapped-ion quantum simulation is able to test basic foundations of quantum mechanics, and to study both ground state properties and dynamics of complicated, long-ranged models.

DECLARATIONS

Acknowledgements

The authors thank Remigiusz Augusiak and Ravindra Chhajlany for discussions. This work has been supported by EU (SIQS, FP7/2007-2013 Grant No. 323714 EQUAM), ERC (QUAGATUA, OSYRIS), Spanish MINCIN (FIS2008-00784 TOQATA), Generalitat de Catalunya (2009-SGR1289), Alexander von Humboldt Stiftung, and Fundació Cellex.

Competing interests

The authors declare that they have no competing interests.

Authors' contributions

All authors contributed equally to the writing of this paper. All authors read and approved the final manuscript.

REFERENCES

1. Auerbach A: Interacting Electrons and Quantum Magnetism. Springer, Berlin; 1994
2. Lewenstein M, Sanpera A, Ahufinger V: Ultracold Atoms in Optical Lattices - Simulating Quantum Many-Body Systems. Oxford University Press, Oxford; 2012.

3. Romero-Isart O, Eckert K, Rodó C, Sanpera A: Transport and entanglement generation in the Bose-Hubbard model. J Phys A, Math Theor 2007., 40(28): Article ID 8019 Article ID 8019

4. Hauke P, Tagliacozzo L: Spread of correlations in long-range interacting quantum systems. Phys Rev Lett 2013., 111: Article ID 207202 Article ID 207202

5. Schachenmayer J, Lanyon BP, Roos CF, Daley AJ: Entanglement growth in quench dynamics with variable range interactions. Phys Rev X 2013., 3: Article ID 031015 Article ID 031015

6. Lamata L, León J, Schätz T, Solano E: Dirac equation and quantum relativistic effects in a single trapped ion. Phys Rev Lett 2007., 98: Article ID 253005 Article ID 253005

7. Gerritsma R, Kirchmair G, Zähringer F, Solano E: Quantum simulation of the Dirac equation. Nature 2010., 463: Article ID 68 Article ID 68

8. Schmied R, Wesenberg JH, Leibfried D: New J Phys. 2011., 13: Article ID 115011 Article ID 115011

9. Casanova J, Mezzacapo A, Lamata L, Solano E: Quantum simulation of interacting fermion lattice models in trapped ions. Phys Rev Lett 2012., 108: Article ID 190502 Article ID 190502

10. Pons M, Ahufinger V, Wunderlich C, Sanpera A, Braungardt S, Sen(De) A, Sen U, Lewenstein M: Trapped ion chain as a neural network: error resistant quantum computation. Phys Rev Lett 2007., 98: Article ID 023003 Article ID 023003

11. Mintert F, Wunderlich C: Ion-trap quantum logic using long-wavelength radiation. Phys Rev Lett 2001., 87: Article ID 257904 Article ID 257904

12. Porras D, Cirac JI: Effective quantum spin systems with trapped ions. Phys Rev Lett 2004., 92: Article ID 207901 Article ID 207901

13. Friedenauer A, Schmitz H, Glueckert JT, Porras D, Schaetz T: Simulating a quantum magnet with trapped ions. Nat Phys 2008, 4(10):757–761. 10.1038/nphys1032

14. Kim K, Chang M-S, Islam R, Korenblit S, Duan L-M, Monroe C: Entanglement and tunable spin-spin couplings between trapped ions using multiple transverse modes. Phys Rev Lett 2009., 103: Article ID 120502 Article ID 120502

15. Kim K, Chang MS, Korenblit S, Islam R, Edwards EE, Freericks JK, Lin GD, Duan LM, Monroe C: Quantum simulation of frustrated Ising spins with trapped ions. Nature 2010, 465: 590–593. 10.1038/nature09071

16. Britton JW, Sawyer BC, Keith AC, Wang C-CJ, Freericks JK, Uys H, Biercuk MJ, Bollinger JJ: Engineered two-dimensional Ising interactions in a trapped-ion quantum simulator with hundreds of spins. Nature 2012, 484(7395):489–492. 10.1038/nature10981

17. Richerme P, Gong Z-X, Lee A, Senko C, Smith J, Foss-Feig M, Michalakis S, Gorshkov AV, Monroe C: arXiv:1401.5088; 2014.

18. Jurcevic P, Lanyon BP, Hauke P, Hempel C, Zoller P, Blatt R, Roos CF: arXiv:1401.5387; 2014.

19. Deng X-L, Porras D, Cirac JI: Effective spin quantum phases in systems of trapped ions. Phys Rev A 2005., 72: Article ID 063407 Article ID 063407

20. Graß T, Juliá-Díaz B, Kuś M, Lewenstein M: Quantum chaos in su(3) models with trapped ions. Phys Rev Lett 2013., 111: Article ID 090404 Article ID 090404

21. Koffel T, Lewenstein M, Tagliacozzo L: Entanglement entropy for the long-range Ising chain in a transverse field. Phys Rev Lett 2012., 109: Article ID 267203 Article ID 267203

22. Haldane FDM:Exact Jastrow-Gutzwiller resonating-valence-bond ground state of the spin-(1/2 antiferromagnetic Heisenberg chain with 1/r2 exchange. Phys Rev Lett 1988, 60: 635–638. 10.1103/PhysRevLett.60.635

23. Shastry BS:Exact solution of an S=1/2 Heisenberg antiferromagnetic chain with long-ranged interactions. Phys Rev Lett 1988, 60: 639–642. 10.1103/PhysRevLett.60.639

24. Greiter M: Mapping of Parent Hamiltonians: from Abelian and Non-Abelian Quantum Hall States to Exact Models of Critical Spin Chains. Springer, Berlin; 2011.

25. Majumdar CK, Ghosh DK: J Math Phys. 1969., 10: Article ID 1388 Article ID 1388

26. Lipkin H, Meshkov N, Glick AJ: Nucl Phys. 1965., 62: Article ID 188 Article ID 188

27. Tura J, Augusiak R, Sainz AB, Vértesi T, Lewenstein M, Acín A: arXiv:1306.6860; 2013.

28. Laughlin RB: Anomalous quantum hall effect: an incompressible quantum fluid with fractionally charged excitations. Phys Rev Lett 1983., 50: Article ID 1395 Article ID 1395
29. Gebhard F, Vollhardt D: Correlation functions for Hubbard-type models: the exact results for the Gutzwiller wave function in one dimension. Phys Rev Lett 1987, 59: 1472–1475. 10.1103/PhysRevLett.59.1472
30. Affleck I: Critical behavior of two-dimensional systems with continuous symmetries. Phys Rev Lett 1985, 55: 1355–1358. 10.1103/PhysRevLett.55.1355
31. Okamoto K, Nomura K:Fluid-dimer critical point in s=12 antiferromagnetic Heisenberg chain with next nearest neighbor interactions. Phys Lett A 1992, 169(6):433–437. 10.1016/0375-9601(92)90823-
32. Eggert S: Numerical evidence for multiplicative logarithmic corrections from marginal operators. Phys Rev B 1996, 54: 9612–9615. 10.1103/PhysRevB.54.R9612
33. Schneider C, Porras D, Schaetz T: Experimental quantum simulations of many-body physics with trapped ions. Rep Prog Phys 2012., 75(2): Article ID 024401 Article ID 024401
34. Korenblit S, Kafri D, Campbell WC, Islam R, Edwards EE, Gong Z-X, Lin G-D, Duan L-M, Kim J, Kim K, Monroe C: Quantum simulation of spin models on an arbitrary lattice with trapped ions. New J Phys 2012., 14(9): Article ID 095024 Article ID 095024
35. Kim K, Korenblit S, Islam R, Edwards EE, Chang M-S, Noh C, Carmichael H, Lin G-D, Duan L-M, Wang CCJ, Freericks JK, Monroe C: Quantum simulation of the transverse Ising model with trapped ions. New J Phys 2011., 13(10): Article ID 105003 Article ID 105003
36. Sandvik AW: Ground states of a frustrated quantum spin chain with long-range interactions. Phys Rev Lett 2010., 104: Article ID 137204 Article ID 137204
37. Chhajlany RW, Tomczak P, Wójcik A, Richter J: Entanglement in the Majumdar-Ghosh model. Phys Rev A 2007., 75: Article ID 032340 Article ID 032340
38. Laflorencie N, Sørensen ES, Chang M-S, Affleck I: Boundary effects in the critical scaling of entanglement entropy in 1d systems. Phys Rev Lett 2006., 96: Article ID 100603 Article ID 100603
39. Them K: arXiv:1308.2882; 2013.
40. Brunner N, Cavalcanti D, Pironio S, Scarani V, Wehner S: Bell nonlocality. Rev. Mod. Phys. 2014, 86: 419. 10.1103/RevModPhys.86.419
41. Bell JS: Physics. 1964., 1: Article ID 195 Article ID 195

42. Douglas JS, Habibian H, Gorshkov AV, Kimble HJ, Chang DE: arXiv:1312.2435; 2013.
43. Noguchi A, Toyoda K, Urabe S: Generation of Dicke states with phonon-mediated multilevel stimulated Raman adiabatic passage. Phys Rev Lett 2012., 109: Article ID 260502 Article ID 260502
44. Blanes S, Casas F, Oteo JA, Ros J: The magnus expansion and some of its applications. Phys Rep 2009, 470(5–6):151–238. 10.1016/j.physrep.2008.11.001

4
CHAPTER
FRACTAL APPROXIMATION OF MOTION AND ITS IMPLICATIONS IN QUANTUM MECHANICS

Maricel Agop[1,2], Daniela Magop[3], Elena Simona Bacaita [1,4]

[1]*Physics Department, Faculty of Machine Manufacturing and Industrial Management, "Gheorghe Asachi" Technical University, Iasi, Romania*

[2]*Lasers, Atoms and Molecules Physics Laboratory, University of Science and Technology, Lille, France*

[3]*Physics Department, "Al. I. Cuza" University, Iasi, Romania*

[4]*Department of Natural and Synthetic Polymers, Faculty of Chemical Engineering and Environmental Protection, "Gheorghe Asachi" Technical University, Iasi, Romania*

ABSTRACT

Inconsistencies of some standard quantum mechanical models (Madelung's, de Broglie's models) are eliminated assuming the micro particle movements on continuous, but non-differentiable curves (fractal curves). This hypothesis, named by us the fractal approximation of motion, will allow an unitary approach of the phenomena in quantum mechanics (separation of the physical motion of objects in wave and particle components depending on the scale of resolution, correlated motions of the wave and particle, i.e. wave-particle duality, the mechanisms of duality, by means of both phase wave-particle coherence and wave-particle incoherence, the particle as a clock, particle incorporation into the wave and the implications of such a process). Moreover, correspondences with standard gravitational models (Einstein's model, string theory) can be also distinguished.

Keywords: Fractals; Quantum Mechanics; Gravitational Models

1. INTRODUCTION

We perform a critical analysis of some quantum mechanical models such as the hydrodynamic model (Madelung's model), de Broglie's theory of double solution etc., specifying both mathematical and physical inconsistencies that occur in their construction.

These inconsistencies are eliminated by means of the fractal approximation of motion (physical objects moving on continuous and non-differentiable curves, i.e. fractal curves) developed in the framework of Scale Relativity (SR) [1-7]. The following original results are obtained: 1) Separation of the physical motion of objects in wave and particle components depending on the scale of resolution (differentiable as waves and non-differentiable as particles)—see Sections 5-7; 2) Correlated motions of the wave and particle (wave-particle duality)—see Section 8; 4) The mechanisms of duality (in phase wave-particle coherence, Sections 9 and 10; and 5) Wave-particle incoherence, see Section 11; 6) The particle as a clock, its incorporation into the wave and the implications of such a process—see Sections 12 and 13.

The original results of this work are published in references [8-12]. Explicitly, Eulerian's approximation of motions on fractal curves is presented in [11,12], the hydrodynamic model in a second order approximation of motion in [8], wave-particle duality for "coherent" fractal fluids with the explanation of the potential gap in [9], the physical self-consistence of wave-particle duality in various approximations of motion and for various fractal curves in [11,12]. A unitary treatment of both the problems listed above and their various mathematical and physical extensions are developed in [10].

2. HYDRODYNAMIC MODEL OF QUANTUM MECHANICS (MADELUNG'S MODEL)

Quantum mechanics is substantiated by the Schrődinger wave equation [13-16]

$$i\hbar\frac{\partial\Psi}{\partial t}=\mathbf{U}\Psi-\frac{\hbar^2}{2m_0}\Delta\Psi \tag{1}$$

where $\hbar$ is the reduced Planck's constant, m_0 the rest mass of the test particle, U the external scalar field and Ψ the wave-function associated to the physical system. This differential equation is linear and complex.

Starting from this equation, Madelung [17,18] constructed the following model. One separates real and imaginary parts by choosing Ψ of the form:

$$\Psi(\mathbf{r},t)=R(\mathbf{r},t)e^{iS(\mathbf{r},t)} \tag{2}$$

which induces the velocity field:

$$v=\frac{\hbar}{m_0}\nabla S \tag{3}$$

and the density of the probability field:

$$\rho(\mathbf{r},t)=R^2(\mathbf{r},t) \tag{4}$$

Using these fields one gets the hydrodynamic version of quantum mechanics (Madelung's model)

$$\frac{\partial}{\partial t}(m_0\rho v)+\nabla(m_0\rho vv)=-\rho\nabla(\mathbf{U}+\mathbf{Q}) \tag{5}$$

$$\frac{\partial\rho}{\partial t}+\nabla\cdot(\rho v)=0 \tag{6}$$

where

$$\mathbf{Q}=-\frac{\hbar^2}{2m_0}\frac{\Delta\sqrt{\rho}}{\sqrt{\rho}} \tag{7}$$

is called the quantum potential. Equation (5) corresponds to the momentum conservation law and Equation (6) to the conservation law of the probability's density field (quantum hydrodynamics equations).

We have the following:

1) Any micro-particle is in constant interaction with an environment called "sub-quantum medium" through the quantum potential Q;

2) The "sub-quantum medium" is identified with a non-relativistic quantum fluid described by the equations of quantum hydrodynamics. In other words, the propagation of the Ψ field from wave mechanics is replaced by a fictitious fluid flow having the density ρ and the speed v, the fluid being in a field of forces $\nabla(\mathbf{U}+\mathbf{Q})$. Moreover, the following model of particle states [19-26], Madelung type fluid in "interaction" with its own "shell" (there is no space limitation of the fluid, though of the particle).

3. DE BROGLIE'S THEORY OF DOUBLE SOLUTION. THE NEED FOR INTRODUCING THE MODEL OF BOHM AND VIGIER

One of the key observations that de Broglie left in the development of quantum mechanics, is the difference between the relativistic transformation of the frequency of a wave and that of a clock's frequency [27-32]. It is well known that, if υ_0 is the frequency of a clock in its own framework, the frequency conferred by an observer who sees it passing with the speed $v = \beta c$ is

$$\nu_c = \nu_0\sqrt{1-\beta^2}$$

This is what is called the phenomenon of "slowing down of horologes". This phenomenon takes place due to the relative motion of horologes. On the contrary, if a wave within a certain reference system is a stationary one, with frequency υ_0 and is noticed in a reference system animated with speed $v = \beta c$, as compared with the first one, it will appear as a progressive wave that propagates in the sense of the relative motion, with frequency

$$\nu = \frac{\nu_0}{\sqrt{1-\beta^2}}$$

and with the phase speed

$$V = \frac{c}{\beta} = \frac{c^2}{v}$$

If the corpuscle, according to relation W = hv, is given an internal frequency

$$\nu_0 = \frac{m_0 c^2}{h}$$

and if we admit that within the appropriate system of the corpuscle the associated wave is a stationary one, with frequency υ_0, all the fundamental relations of wave mechanics and in particular $\lambda = h/p$, in which p is the impulse of the corpuscle, are immediately obtained from the previous relations.

Since de Broglie considers that the corpuscle is constantly located in the wave, he notices the following consequence: the motion of the corpuscle has such a nature that it ensures the permanent concordance between the phase of the surrounding wave and the internal phase of the corpuscle considered as a small horologe. This relation can be immediately verified in the simple case of a corpuscle in uniform motion, accompanied by a monochromatic plain wave. Thus, when the wave has the general form

$$\Psi = \mathbf{A}(x,y,z,t)e^{\frac{2\pi i}{h}\Phi(x,y,z,t)}$$

in which A and Φ are real, the phase concordance between the corpuscle and its wave requires that the speed of the corpuscle in each point of its trajectory be given by the relation

$$v = -\frac{1}{m_0}\nabla\Phi$$

Nevertheless it was not enough to superpose the corpuscle with the wave, imposing it to be guided by the propagation of the wave: the corpuscle had to be represented as being incorporated in the wave, i.e. as being a part of the structure of the wave. De Broglie was thus directed to what he himself called the theory of "double solution". This theory admits that the real wave is not a homogeneous one, that it has a very small area of high concentration of the field that represents the corpuscle and that, besides this very small area, the wave appreciably coincides with the homogeneous wave as formulated by the usual wave mechanics.

The phenomenon of guiding the particle by the surrounding wave field results from the fact that the equations of the field are not linear ones and that this lack of linearity, that almost exclusively shows itself in the corpuscular area, correlating the motion of the particle with the propagation of the surrounding wave [30-33].

Nevertheless there is a consequence of "guidance" upon which we should insist. Even if a particle is not submitted to any external field, if the wave that surrounds it is not an appreciably plain and monochromatic one (therefore if this wave has to be represented through a superposition of monochromatic plain waves) the motion that the guidance formula imposes is not rectilinear and uniform. The corpuscle is subjected by the surrounding wave, to a force that curves its trajectory: this "quantum force" equals the gradient with the changed sign of the quantum potential Q given by (7). Therefore, the uniform motion of the wave has to be superposed with a "Brownian" motion having random character that is specific to the corpuscle.

Under the influence of Q, the corpuscle, instead of uniformly following one of the trajectories that are defined by the guidance law, constantly jumps from one of these trajectories to another, thus passing in a very short period of time, a considerably big number of sections within these trajectories and, while the wave remains isolated in a finite area of the space, this zigzag trajectory hurries to explore completely all this region. In this manner, one can justify that the probability of the particle to be present in a volume element $d\tau$ of the physical space is equal to $|\Psi|^2 d\tau$. This is what Bohm and Vigier did in their statement: therefore they showed that the probability of repartition in $|\Psi|^2$ must take place very quickly. The success of this demonstration must be correlated with the characteristics if "Markov's chains" [19,34-36].

4. COMMENTS

In his attempt to built the theory of the double solution, de Broglie admits certain assertions [27-32]: 1) The frequency of the corpuscle that is assimilated to a small horologe must be identified with the frequency of the associated progressive wave; 2) The coherence of the inner phase of the corpuscle-horologe with the phase of the associated wave; 3) The corpuscle must be "incorporated" into the progressive associated wave through the "singularity" state. Thus, the motion of the corpuscle "correlates" with the propagation of the associated progressive wave. Nevertheless, once we admit these statements, de Broglie's theory does not answer a series of problems, such as, for example: 1) What are the mechanisms through which either the wave feature or the corpuscular one impose, either both of them in the stationary case as well as in the non stationary one? 2) The limits in the wave-corpuscle system of the corpuscular component as well as the wave one and their correspondence; 3) How is the "solidarity" between the motion of the corpuscle and the one of the associated progressive wave naturally induced? 4) What are the consequences of this "solidarity"? Others discrepancies are mentioned in [27- 32].

Moreover, Madelung's theory [17,18] brings new problems. How can we built a pattern of a corpuscle (framework + Madelung liquid) endlessly extended in space?

Here are some of the "drawbacks" of the patterns in Sections 2 and 3 which we shall analyze and remove by means of introducing the fractal approximation of the motion.

5. THE MOTION EQUATION OF THE PHYSICAL OBJECT IN THE FRACTAL APPROXIMATION OF MOTION. THE EULERIAN SEPARATION OF MOTION ON RESOLUTION SCALES

The fractal approximation of motion refers to the movement of physical objects (wave + corpuscle) on continuous and non differentiable curves (fractal curves). This approximation is based on the scale Relativity theory (RS) [1-7]. Thus, the fractal differential operator can be introduced

$$\frac{\hat{\mathrm{d}}}{\mathrm{d}t} = \frac{\partial}{\partial t} + \hat{\mathbf{V}} \cdot \nabla - i\frac{\lambda^2}{2\tau}\left(\frac{\mathrm{d}t}{\tau}\right)^{(2/D_F)-1} \tag{8}$$

where $\hat{\mathbf{V}}$ is the complex speed field

$$\hat{\mathbf{V}} = \mathbf{V} - i\mathbf{U} \tag{9}$$

λ is the scale length, dt is the temporary resolution scale, τ is the specific time to fractal-non fractal transition, and D_F is the arbitrary and constant fractal dimension. Regarding the fractal dimension, we can use any of Hausdorff-Bezicovici, Minkowski-Bouligand or Kolmogoroff dimensions, etc. [3,5-7,36-46]. The only restriction refers to the maintaining of the same type of fractal dimension during the dynamic analysis. The real part of the speed field $\mathbf{V}$ is differentiable and independent as compared with the resolution scale, while the imaginary scale $\mathbf{U}$ is non differentiable (fractal) and depends on the resolution scale.

Now we can apply the principle of scale covariance by substituting the standard time derivate $(\mathrm{d}/\mathrm{d}t)$ with the complex operator $\hat{\mathrm{d}}/\mathrm{d}t$. Accordingly, the equation of fractal space-time geodesics (the motion equation in second order approximation, where second order derivates are used) in a covariant form:

$$\frac{\hat{\mathrm{d}}\hat{\mathbf{V}}}{\mathrm{d}t}=\frac{\partial\hat{\mathbf{V}}}{\partial t}+\hat{\mathbf{V}}\cdot\nabla\hat{\mathbf{V}}-i\frac{\lambda^2}{2\tau}\left(\frac{\mathrm{d}t}{\tau}\right)^{(2/D_F)-1}\nabla^2\hat{\mathbf{V}}\equiv 0 \tag{10}$$

This means that the sum of the local acceleration $\partial\hat{\mathbf{V}}/\partial t$, convection $\hat{\mathbf{V}}\cdot\nabla\hat{\mathbf{V}}$ and "dissipation" $\nabla^2\hat{\mathbf{V}}$ reciprocally compensate in any point of the arbitrarily fractal chosen trajectory of a physical object.

Formally, (10) is a Navier-Stokes type equation, with an imaginary viscosity coefficient,

$$\eta=i\frac{\lambda^2}{2\tau}\left(\frac{\mathrm{d}t}{\tau}\right)^{(2/D_F)-1} \tag{11}$$

This coefficient depends on two temporary scales, as well as on a length scale. The existence of a pure imaginary structured coefficient specifies the fact that "the environment" has rheological features (viscoelastic and hysteretic ones [47-49].

For

$$\frac{\lambda^2}{2\tau}\times\left(\frac{\mathrm{d}t}{\tau}\right)^{(2/D_F)-1}\to 0 \tag{12}$$

Equation (10) reduces to Euclidian form [10-12]:

$$\frac{\partial \hat{\mathbf{V}}}{\partial t}+\hat{\mathbf{V}}\cdot\nabla\hat{\mathbf{V}}\equiv 0 \tag{13}$$

and, hence, separating the real part from the imaginary one

$$\frac{\partial \mathbf{V}}{\partial t}+\mathbf{V}\cdot\nabla\mathbf{V}-\mathbf{U}\cdot\nabla\mathbf{U}=0 \tag{14a}$$

$$\frac{\partial \mathbf{U}}{\partial t}+\mathbf{U}\cdot\nabla\mathbf{V}+\mathbf{V}\cdot\nabla\mathbf{U}=0 \tag{14b}$$

Equation (14a) corresponds to the law of the impulse conservation at differentiable scale (the undulatory component), while (14b) corresponds to the same law, but at a non differentiable scale (corpuscular component). As we will later show, in the case of irrotational movements (14) it will be assimilated to the law of mass conservation.

6. ROTATIONAL MOTIONS AND FLOW REGIMES OF A FRACTAL FLUID

For rotational motions, $\nabla\times\hat{\mathbf{V}}\neq 0$ relation (10) with (9) through separating the real part from the imaginary one, i.e. through separating the motions at a differential scale (undulatory characteristic) and non differential one (corpuscular characteristic), results [10].

$$\frac{\partial \mathbf{V}}{\partial t}+\mathbf{V}\cdot\nabla\mathbf{V}-\mathbf{U}\cdot\nabla\mathbf{U}-\frac{\lambda^2}{2\tau}\left(\frac{dt}{\tau}\right)^{(2/D_F)-1}\Delta\mathbf{U}=0 \tag{15a}$$

$$\frac{\partial \mathbf{U}}{\partial t}+\mathbf{U}\cdot\nabla\mathbf{V}+\mathbf{V}\cdot\nabla\mathbf{U}+\frac{\lambda^2}{2\tau}\left(\frac{dt}{\tau}\right)^{(2/D_F)-1}\Delta\mathbf{V}=0 \tag{15b}$$

According to the operator relations

$$\mathbf{V}\cdot\nabla\mathbf{V}=\nabla\left(\frac{\mathbf{V}^2}{2}\right)-\mathbf{V}\times(\nabla\times\mathbf{V}) \tag{16a}$$

$$\mathbf{U}\cdot\nabla\mathbf{U}=\nabla\left(\frac{\mathbf{U}^2}{2}\right)-\mathbf{U}\times(\nabla\times\mathbf{U}) \tag{16b}$$

$$\mathbf{U}\cdot\nabla\mathbf{V}+\mathbf{V}\cdot\nabla\mathbf{U}=\nabla(\mathbf{U}\cdot\mathbf{V})-\mathbf{V}\times(\nabla\times\mathbf{U})-\mathbf{U}\times(\nabla\times\mathbf{V}) \tag{16c}$$

Equations (15) take equivalent forms

$$\frac{\partial\mathbf{V}}{\partial t}+\nabla\left(\frac{\mathbf{V}^2}{2}-\frac{\mathbf{U}^2}{2}\right)-\mathbf{V}\times(\nabla\times\mathbf{V})$$
$$-\mathbf{U}\times(\nabla\times\mathbf{U})-\frac{\lambda^2}{2\tau}\left(\frac{\mathrm{d}t}{\tau}\right)^{(2/D_F)-1}\Delta\mathbf{U}=0 \tag{17a}$$

$$\frac{\partial\mathbf{U}}{\partial t}+\nabla(\mathbf{V}\cdot\mathbf{U})-\mathbf{V}\times(\nabla\times\mathbf{U})$$
$$-\mathbf{U}\times(\nabla\times\mathbf{V})+\frac{\lambda^2}{2\tau}\left(\frac{\mathrm{d}t}{\tau}\right)^{(2/D_F)-1}\Delta\mathbf{V}=0 \tag{17b}$$

We can now characterize the flow regimes of the fractal fluid at different scales, using some classes of Reynolds numbers. At a differential scale we have

$$R(\text{differential}-\text{nondifferential})=R(D-N)$$
$$=\frac{|\mathbf{V}\cdot\nabla\mathbf{V}|}{D|\nabla\mathbf{U}|}\approx\frac{\mathbf{V}^2l^2}{D\mathbf{U}L} \tag{18}$$

$$R(\text{differential}-\text{nondifferential})=R(D-N)$$
$$=\frac{|\mathbf{U}\cdot\nabla\mathbf{U}|}{D|\nabla\mathbf{U}|}\approx\frac{\mathbf{U}l}{D} \tag{19}$$

with

$$D=\frac{\lambda^2}{2\tau}\left(\frac{\mathrm{d}t}{\tau}\right)^{(2/D_F)-1} \tag{20}$$

and at nondifferential scale.

$$R(\text{differential}-\text{nondifferential}-\text{differential transition})$$
$$=R(\mathrm{TDN}-D)=\frac{|\mathbf{U}\cdot\nabla\mathbf{V}|}{D|\Delta\mathbf{V}|}\approx\frac{\mathbf{U}l}{D} \tag{21}$$

$$R(\text{differential}-\text{nondifferential}-\text{differential transition})$$
$$=R(\text{TDN}-D)=\frac{|\mathbf{V}\cdot\nabla\mathbf{U}|}{D|\Delta\mathbf{V}|}\approx\frac{\mathbf{U}L^2}{Dl} \tag{22}$$

In previous relations V, L, D, are the specific parameters, while U, l, D are the parameters of the nondifferential scale. The parameters V, U are specific speeds, L, l specific lengths and D is a viscosity coefficient. Moreover, the common "element" for R(D ¯N), R(N ¯N), R(TDN ¯D) and R(TND ¯D) is the "viscosity" which, through (20) is imposed by the resolution scale.

Equations (15) are simplified in the case of the stationary motion for small Reynolds numbers. Thus, Equation (15) for small R(D ¯N) becomes

$$-\mathbf{U}\cdot\nabla\mathbf{U}-\frac{\lambda^2}{2\tau}\left(\frac{dt}{\tau}\right)^{(2/D_F)-1}\nabla\mathbf{U}=0 \tag{23}$$

and for small R(N ¯N)

$$-\mathbf{V}\cdot\nabla\mathbf{V}-\frac{\lambda^2}{2\tau}\left(\frac{dt}{\tau}\right)^{(2/D_F)-1}\nabla\mathbf{U}=0 \tag{24}$$

Equation (15b) for small R(TDN ¯D) takes the form

$$\mathbf{V}\cdot\nabla\mathbf{U}+\frac{\lambda^2}{2\tau}\left(\frac{dt}{\tau}\right)^{(2/D_F)-1}\Delta\mathbf{V}=0 \tag{25}$$

and for small R(TND ¯D)

$$\mathbf{U}\cdot\nabla\mathbf{V}+\frac{\lambda^2}{2\tau}\left(\frac{dt}{\tau}\right)^{(2/D_F)-1}\Delta\mathbf{V}=0 \tag{26}$$

7. Irotational Motions of a Fractal Fluid. The Incorporation of the Associate Wave Corpuscle through the Solidarity of Movements and Generation of Schrodinger Equation

For irrotational motions

$$\nabla\times\hat{\mathbf{V}}=0 \tag{27}$$

which implies

$$\nabla\times\mathbf{V}=0,\nabla\times\mathbf{U}=0 \tag{28}$$

$$\frac{\partial \mathbf{V}}{\partial t}+\nabla\left(\frac{\mathbf{V}^2}{2}\right)-i\frac{\lambda^2}{2\tau}\left(\frac{dt}{\tau}\right)^{(2/D_F)-1}\Delta\mathbf{V}=0 \tag{29}$$

Since through (27) the complex speed field is expressed by means of a scalar function gradient F,

$$\hat{\mathbf{V}}=\nabla\Phi \tag{30}$$

Equation (29) taking into account the operator identities

$$\frac{\partial}{\partial t}\nabla=\nabla\frac{\partial}{\partial t},\quad \nabla\Delta=\Delta\nabla \tag{31}$$

takes the form

$$\nabla\left[\frac{\partial\Phi}{\partial t}+\frac{1}{2}(\nabla\Phi)^2-\mathrm{i}\frac{\lambda^2}{2\tau}\left(\frac{dt}{\tau}\right)^{(2/D_F)-1}\Delta\Phi\right]=0 \tag{32}$$

or furthermore, through integration

$$\frac{\partial\Phi}{\partial t}+\frac{1}{2}(\nabla\Phi)^2-i\frac{\lambda^2}{2\tau}\left(\frac{dt}{\tau}\right)^{(2/D_F)-1}\Delta\Phi=F(t) \tag{33}$$

where F(t) is an arbitrary function depending only on time.

In particular, for F having the form

$$\Phi=-2i\frac{\lambda^2}{2\tau}\left(\frac{dt}{\tau}\right)^{(2/D_F)-1}\ln\Psi \tag{34}$$

where Y is a new complex scalar function, Equation (46), with the operator identity

$$\frac{\Delta\Psi}{\Psi}=\Delta\ln\Psi+(\nabla\ln\Psi)^2 \tag{35}$$

takes the form:

$$\frac{\lambda^4}{4\tau^2}\left(\frac{dt}{\tau}\right)^{(4/D_F)-2}\Delta\Psi+i\frac{\lambda^2}{2\tau}\left(\frac{dt}{\tau}\right)^{(2/D_F)-1}\frac{\partial\Psi}{\partial t}+\frac{F(t)}{2}\Psi=0 \tag{36}$$

The Schrodinger “geodesics” can be obtained as a particular case of Equation (36), based on the following hypothesis (conditions of solidarity of the motion, incorporating the associated wave corpuscle): 1) The motions of the micro-particles take place on fractal curves with the fractal dimension D_F = 2, i.e. the Peano curves [3,5]; 2) $d_{\pm}\xi^i$ are the Markov-Wiener type stochastic variables [3,5] that satisfy the rule

$$\left\langle d_{\pm}\xi^i d_{\pm}\xi^l \right\rangle = \pm\delta^{il}\frac{\lambda^2}{\tau}\mathrm{d}t \tag{37}$$

3) Space scale l and temporary one τ are specific for the Compton scale

$$\lambda = \frac{\hbar}{m_0 c}, \quad \tau = \frac{\hbar}{m_0 c^2} \tag{38}$$

with m_0 the rest mass of the microparticle, c the speed of light in vacuum and $\hbar$ the reduced Planck constant. The parameters (38) should not be understood as "structures" of the standard space-time, but as standards of scale space-time; 4) Function F(t) from (36) is null. Under these circumstances, (36) is reduced to the standard form of Schrodinger's Equation [13,15]

$$\frac{\hbar^2}{2m_0}\Delta\Psi + i\hbar\frac{\partial\Psi}{\partial t} = 0 \tag{39}$$

In such a context, the scale potential of the complex speeds plays the role of the wave function.

8. Extended Hydrodynamic Model of Scale Relativity and Incorporation of Associated Wave Corpuscle through Fractal Potential. The Correspondence with Madelung Model

Substituting the complex speed (9) with the restriction (27) and separating the real part with the imaginary one, we obtain the set of differential Equations [10]

$$m_0\frac{\partial \mathbf{V}}{\partial t} + m_0\nabla\left(\frac{\mathbf{V}^2}{2}\right) = -\nabla(\mathbf{Q}) \tag{40a}$$

$$\frac{\partial \mathrm{U}}{\partial t} + \nabla(\mathbf{V}\cdot\mathbf{U}) + \frac{\lambda^2}{2\tau}\left(\frac{\mathrm{d}t}{\tau}\right)^{(2/D_F)-1}\Delta\mathbf{V} = 0 \tag{40b}$$

where Q is the fractal potential, expressed as follows:

$$\mathbf{Q} = -\frac{m_0 \mathbf{U}^2}{2} - \frac{m_0}{2}\frac{\lambda^2}{2\tau}\left(\frac{dt}{\tau}\right)^{(2/D_F)-1} \nabla \cdot \mathbf{U} \tag{41}$$

For

$$\Psi = \sqrt{\rho e}^{iS} \tag{42}$$

with $\sqrt{\rho}$ an amplitude and S a phase, then (34) under the form

$$\Phi = -i\frac{\lambda^2}{\tau}\left(\frac{dt}{\tau}\right)^{(2/D_F)-1} \ln\left(\sqrt{\rho e}^{iS}\right)$$

implies the complex speed fields of components

$$\mathrm{V} = \frac{\lambda^2}{2\tau}\left(\frac{dt}{\tau}\right)^{(2/D_F)-1} \nabla S \tag{43a}$$

$$\mathbf{U} = \frac{\lambda^2}{2\tau}\left(\frac{dt}{\tau}\right)^{(2/D_F)-1} \nabla \ln \rho \tag{43b}$$

From the perspective of Equation (43), the Equation (40) keeps its form, and the fractal potential is given by the simple expression

$$\mathbf{Q} = -m_0 \frac{\lambda^2}{\tau}\left(\frac{dt}{\tau}\right)^{(2/D_F)-1} \frac{\Delta\sqrt{\rho}}{\sqrt{\rho}} \tag{44}$$

Again through Equations (43), Equation (40b) takes the form:

$$\nabla\left(\frac{\partial \ln \rho}{\partial t} + \mathbf{V}\cdot\nabla \ln \rho + \nabla\cdot\mathbf{V}\right) = 0$$

or, still, through integration with $\rho \neq 0$

$$\frac{\partial \rho}{\partial t} + \nabla\cdot(\rho\mathbf{V}) = T(t) \tag{45}$$

with T(t), an exclusively time dependent function Equation (40) corresponds to the impulse conservation law at differential scale (the classical one), while the impulse conservation law at non differential scale is expressed through (45) with $T(t)\equiv 0$, as a probability density conservation law.

Therefore, equations

$$m_0\left(\frac{\partial \mathbf{V}}{\partial t}+\mathbf{V}\cdot\nabla\left(\frac{\mathbf{V}^2}{2}\right)\right)=-\nabla(\mathbf{Q}) \tag{46a}$$

$$\frac{\partial \rho}{\partial t}+\nabla\cdot(\rho\mathbf{V})=0 \tag{46b}$$

with Q given by (41) or (44) forms the set of equations of scale relativity extended hydrodynamics in fractal dimension D_F. We mention that in references [3,5-7] the model has been extended only for $D_F = 2$. The fractal potential (41) or (44) is induced by the non differentiability of space-time.

In an external scalar field U, the system of Equation (46) modifies as follows:

$$m_0\left[\frac{\partial \mathbf{V}}{\partial t}+\nabla\left(\frac{\mathbf{V}^2}{2}\right)\right]=-\nabla(\mathbf{Q}+\mathbf{U}) \tag{47a}$$

$$\frac{\partial \rho}{\partial t}+\nabla(\rho\mathbf{V})=0 \tag{47b}$$

Now the quantum mechanics in hydrodynamic formula (Madelung's model [17]) is obtained as a particular case of relations (47), using the following hypothesis: 1) The motion of the micro-particles takes place on Peano curves with $D_F = 2$; 2) $d_\pm \xi^i$ are the Markov-Wiener variables [3,5-7]; 3) The time space scale is a Compton one. Then, (38) have the expressions

$$\mathbf{V}=\frac{\hbar}{m_0}\nabla S,\quad \mathbf{U}=\frac{\hbar}{2m_0}\nabla \ln\rho \tag{48}$$

and (41),

$$\mathbf{Q}=-\frac{m_0\mathbf{U}^2}{2}-\frac{\hbar}{2}\nabla\cdot\mathbf{U} \tag{49}$$

9. "Mechanisms" of Duality through Coherence in Corpuscle-Wave Phase

In the stationary case, the system of Equations (46(a) and (b)) become [10]

$$\nabla\left(\frac{\mathbf{V}^2}{2}+\mathbf{Q}\right)=0 \tag{50a}$$

$$\nabla(\rho\mathbf{V})=0 \tag{50b}$$

or, still, through integration

$$\frac{\mathbf{V}^2}{2}+\mathbf{Q}=E=\text{const.} \tag{51a}$$

$$\rho\mathbf{V}=\text{const.} \tag{51b}$$

Let us choose the null power density in (51b). Then there is no impulse transport at differential scale between corpuscle and wave. Moreover, for $\rho \neq 0$.

$$\mathbf{V}=0 \tag{52}$$

which implies through relation (43)

$$S=\text{const.} \tag{53}$$

In other words, the fluid becomes coherent (the fluid particles have the same phase). Such a state is specific for quantum fluids [50,51], such as superconductors, super-fluids, etc. [14,52]. Under such circumstances, the phase of the corpuscle considered as a small horologe equals the phase of the associated wave (coherence in corpuscle-wave phase).

At non-differential scale, Equation (51), with restriction (52) takes the form

$$\mathbf{Q}=-\frac{2m_0D^2\Delta\sqrt{\rho}}{\sqrt{\rho}}=-\frac{m_0\mathbf{U}^2}{2}-m_0D\nabla\mathbf{U}=E=\text{const} \tag{54a}$$

$$\mathbf{D}=\frac{\lambda^2}{\tau}\left(\frac{dt}{\tau}\right)^{(2/D_F)-1} \tag{54b}$$

or, still, by applying the gradient operator

$$\mathbf{A}=\nabla\left(\sqrt{\rho}\right) \tag{55}$$

$$\Delta\mathbf{A}+\frac{E}{2m_0D^2}\mathbf{A}=0 \tag{56}$$

We distinguish the following situations:

1) For E > 0 and with substitution

$$\frac{1}{\Lambda^2} = \frac{E}{2m_0 D^2} \tag{57}$$

Equation (56) becomes

$$\Delta\mathbf{A} + \frac{1}{\Lambda^2}\mathbf{A} = 0 \tag{58}$$

Therefore: a) The space oscillations of field A and, therefore the space associated with the motion of coherent fluid particles is endowed with regular non homogeneities (of lattice type). In other words, the field A crystallizes the space. The one dimensional space "crystal" has the constant of the network

$$\Lambda = \frac{\lambda^2}{2\tau}\left(\frac{\mathrm{d}t}{\tau}\right)^{(2/D_F)-1}\left(\frac{2m_0}{E}\right)^{1/2} \tag{59}$$

that depends both on the "viscosity" – iη given by (11) and on the energy of the particle; b) The one dimensional geodesics of the "crystallized" space given by the expression

$$\rho(x) = \mathbf{A}^2 \sin^2(kx + \delta) \tag{60}$$

implies both fractal speed

$$\mathbf{U}_x = D\frac{\mathrm{d}\ln\rho}{\mathrm{d}x} = 2Dkc\,\mathrm{tg}(kx + \delta) \tag{61}$$

and fractal potential

$$\mathbf{Q}_x = -\frac{m_0\mathbf{U}_x^2}{2} - m_0 D\frac{\mathrm{d}\mathbf{U}_x}{\mathrm{d}x} = -2m_0 D^2 k^2 \mathrm{ctg}^2(kx + \delta) + 2m_0 D^2 k^2 \frac{1}{\sin^2(kx + \delta)} = 2m_0 D^2 k^2 \tag{62}$$

with A and δ, the integration constants

$$k = \frac{1}{\Lambda} \tag{63}$$

c) For the movements of micro particles on Peano curves (D_F = 2) at Compton scale

$$D = 2m_0\hbar$$

Therefore, through (62), under the form

$$\mathbf{Q}_x = 2m_0 D^2 k^2 = \frac{p_x^2}{2m_0}, \tag{64a}$$

$$p_x = 2m_0 Dk \tag{64b}$$

De Broglie "quantum" impulse is found

$$p_x = \frac{\hbar}{\Lambda} \tag{65}$$

d) The dominant of the wave characteristic is achieved by the "self diffraction" mechanism of the fractal field, ρ, on the one dimensional space "crystal" of constant Λ induced by the same field. Indeed, relation (61) with notations

$$\Phi = kx + \delta, \tag{66a}$$

$$k = \frac{1}{\Lambda} \tag{66b}$$

In approximation $\Phi \approx 1$, i.e. for $\mathrm{tg}\Phi \approx \sin\Phi$ and using Nottale's relation [1-7] $2D/\mathbf{U}_x \approx n\lambda$ it takes the common form (Bragg's relation)

$$\Lambda \sin\Phi \approx n\lambda \tag{67}$$

This result is in concordance with the recently expressed opinion in [53-55]; e) There is impulse transfer on the fractal field between the corpuscle and the wave; f) According to Taylor's criterion [56] self-organization (crystallization and self diffraction of the space) appears when the energy of the system is minimal. This can be immediately verified using relation (51a).

2) For E = 0, Equations (51a) and (56) have the same form

$$\Delta\sqrt{\rho}=0 \quad \Delta\mathbf{A}=0 \tag{68}$$

It follows that: a) The geodesics are expressed through harmonic functions and the particle finds itself in a critical state, i.e. the one that corresponds to the wave-corpuscle transition; b) In the one dimensional case, the geodesics have the form

$$\rho(x)=kx+\delta \tag{69}$$

which induces the fractal speed field

$$\mathbf{U}_x=\frac{D}{kx+\delta} \tag{70}$$

namely the null value of the fractal potential

$$\mathbf{Q}_x=-\frac{m_0}{2}\frac{D^2}{(kx+\delta)^2}+\frac{m_0}{2}\frac{D^2}{(kx+\delta)^2}=0 \tag{71}$$

c) Although the energy is null, there is impulse transfer between corpuscle and wave on the fractal component of the speed field.

3) For E < 0 and with notations

$$\frac{1}{\overline{\Lambda}^2}=\frac{\overline{E}}{2m_0D^2}, \quad E=-\overline{E} \tag{72}$$

Equation (56) takes the form

$$\Delta\mathbf{A}-\frac{1}{\overline{\Lambda}^2}\mathbf{A}=0 \tag{73}$$

The following aspects result: a) Field A is expelled from the structure, its penetration depth being

$$\overline{\Lambda}=\frac{\lambda^2}{2\tau}\left(\frac{\mathrm{d}t}{\tau}\right)^{(2/D_F)-1}\left(\frac{2m_0}{\overline{E}}\right)^{1/2} \quad (74)$$

b) The one-dimensional geodesics of the space are described through function

$$\rho(x)=\overline{\mathrm{A}}^2\mathrm{sh}^2\left(\overline{k}x+\overline{\delta}\right) \quad (75)$$

and lead to the fractal speed

$$\mathbf{U}_x=2D\overline{k}cth\left(\overline{k}x+\overline{\delta}\right) \quad (76)$$

the fractal potential respectively

$$\mathbf{Q}_x=-2m_0D^2\overline{k}^2\mathrm{cth}^2\left(\overline{k}x+\overline{\delta}\right)+2m_0D^2\frac{\overline{k}^2}{\mathrm{sh}^2\left(\overline{k}x+\overline{\delta}\right)}=-2m_0D^2\overline{k}^2 \quad (77)$$

where $\overline{\mathbf{A}},\overline{\delta}$ are two integration constants

$$\overline{k}=\frac{1}{\overline{\Lambda}} \quad (78)$$

c) The dominant of the corpuscular characteristic is accomplished by means of "self-expulsion" mechanism of the fractal field from its own structure that it generates (that is the corpuscle), the penetration depth being $\overline{\Lambda}$. The identification

$$\mathbf{Q}_x=-2m_0D^2\overline{k}^2=\frac{\overline{p}^2}{2m_0} \quad (79)$$

implies the purely imaginary impulse

$$\overline{p}=-2im_0D\overline{k} \quad (80)$$

that suggests ultra rapid virtual states (ultra rapid motions in the wave field, resulting in the "singularity" of the field, i.e. the corpuscle). As a matter of fact, if we consider de Broglie's original theory (motions on Peano curves with $D_F = 2$, at Compton's scale), singularity (the corpuscle) moves "suddenly" and chaotically in the wave field, the wave-corpuscle coupling being accomplished through the fractal potential. The corpuscle "tunnels" the potential barrier imposed by the field of the associate progressive wave, generating particle-antiparticle type pairs (ghost type fields (Bittner E.R., 2000)). Nevertheless this model cannot specify the type of the physical process by means of which we reach such a situation: it is only the second quantification that can do this [50, 51,54-57]; d) There is an impulse transfer between the corpuscle and the wave on the fractal component of the speed field, so that all the attributes of the differential speed could be transferred on the fractal speed.

All the above results indicate that wave-particle duality is an intrinsic property of space and not of the particle.

10. WAVE-CORPUSCLE DUALITY THROUGH FLOWING STATIONARY REGIMES OF A COHERENT FRACTAL FLUID IN PHASE. THE POTENTIAL WELL

According to the previous paragraph, let us study the particle in a potential well with infinite width and walls. Then the speed complex field has the form [8-12]

$$\hat{\mathbf{V}}_x = \mathbf{V}_x - i\mathbf{U}_x = 0 - 2iD\left(\frac{n\pi}{a}\right)\mathrm{ctg}\left(\frac{n\pi}{a}\right)x \tag{81}$$

and generates the fractal potential (the energy of the structure) under the form of the noticeable

$$\mathbf{Q}_n = 2m_0 D^2\left(\frac{n\pi}{a}\right)^2 = E_n \tag{82}$$

The last relation (82) allows the implementation of Reynold's criterion

$$R(n)=\frac{\mathbf{V}_c a}{D}=2n\pi,\quad \mathbf{V}_c=\left(\frac{2E_n}{m_0}\right)^{\frac{1}{2}} \tag{83}$$

For movements on Peano curves (D_F = 2) at Compton scale $\left(2m_0 D=\hbar\right)$ (83) with substitutions

$$m_0\mathbf{V}_c=\Delta P_x,\quad a=\Delta x \tag{84}$$

and n = 1 reduces to Heinsenberg's relation of uncertainty under equal form

$$\Delta p_x \Delta x=\frac{h}{2} \tag{85}$$

while for $n<+\infty$ it implies a Ruelle-Takens' type criterion of evolution towards chaos [58,59]. Therefore, the wave-corpuscle duality is accomplished through the flowing regimes of a fractal fluid that is coherent in phase. Thus, the laminar flow (small n) induces a dominant ondulatory characteristic, while the turbulent flow (big n) induces a dominant corpuscular characteristic.

11. WAVE-CORPUSCLE DUALITY THROUGH NON-STATIONARY REGIMES OF AN INCOHERENT FRACTAL FLUID

In the one dimensional case the equations of hydrodynamics (46) take the form

$$m_0\left(\frac{\partial \mathbf{V}}{\partial t}+\mathbf{V}\cdot\frac{\partial \mathbf{V}}{\partial x}\right)=-\frac{\partial}{\partial x}\left[-2m_0 D^2\frac{1}{\rho^{1/2}}\frac{\partial^2}{\partial x^2}\left(\rho^{1/2}\right)\right] \tag{86a}$$

$$\frac{\partial \rho}{\partial t}+\frac{\partial}{\partial x}\left(\rho\mathbf{V}\right)=0 \tag{86b}$$

Imposing the initial conditions

$$\mathbf{V}\left(x,t=0\right)=c=\text{const} \tag{87a}$$

$$\rho(x,t=0)=\frac{1}{\pi^{1/2}\alpha}e^{-\left(\frac{x}{\alpha}\right)^2}=\rho_0 \tag{87b}$$

and on the frontier

$$\mathbf{V}(x=ct,t)=c; \tag{88a}$$

$$\rho(x=-\infty,t)=\rho(x=+\infty,t)=0 \tag{88b}$$

The solutions of the system (86), using the method in [60], have the expressions

$$\rho(x,t)=\frac{1}{\pi^{1/2}\left[\alpha^2+\left(\frac{2D}{\alpha}t\right)^2\right]^{1/2}}\exp\left[\frac{(x-ct)^2}{\alpha^2+\left(\frac{2D}{\alpha}t\right)^2}\right] \tag{89a}$$

$$\mathbf{V}=\frac{c\alpha^2+\left(\frac{2D}{\alpha}\right)^2 tx}{\alpha^2+\left(\frac{2D}{\alpha}t\right)^2} \tag{89b}$$

The complex speed field is obtained

$$\hat{\mathbf{V}}=\mathbf{V}-i\mathbf{U}=\frac{c\alpha^2+\left(\frac{2D}{\alpha}\right)^2 tx}{\alpha^2+\left(\frac{2D}{\alpha}t\right)^2}+2iD\frac{x-ct}{\alpha^2+\left(\frac{2D}{\alpha}t\right)^2} \tag{90}$$

and the field of fractal forces

$$F=4m_0D^2\frac{(x-ct)}{\left[\alpha^2+\left(\frac{2D}{\alpha}t\right)^2\right]^2} \tag{91}$$

Therefore: 1) Both differential scale speed V and nondifferential one U are not homogeneous in x and t. Under the action of fractal force F, the corpuscle is assimilated to the wave, is a part of its structure, so that it joins the movement of the corpuscle with the propagation of the associated progressive wave; 2) The timing of the movements at the two scales, V = U implies the space-time homographic dependence

$$x = \frac{c\alpha^2}{2D} \frac{1 + \frac{2D}{\alpha^2} t}{1 - \frac{2D}{\alpha^2} t} \tag{92}$$

in the field of forces

$$F = \frac{2m_0 Dc}{\left(1 - \frac{2D}{\alpha^2} t\right)\left[\alpha^2 + \left(\frac{2D}{\alpha} t\right)^2\right]} \tag{93}$$

Considering that the type (92) changes are implied in gravitational interaction [61,62], it follows that the solidarity of the corpuscle movement with the movement of the associated progressive wave is accomplished by means of the appropriate gravitational field of the physical object; 3) The uniform movement V = c is obtained for null fractal force F = 0 and fractal speed U = 0, using condition x = ct. The fractal forces in the semi space. $-\infty \leq x \leq \overline{x}$ and $\overline{x} \leq x \leq +\infty$ are reciprocally compensated.

$$F\Big|_{-\infty}^{\overline{x}} = F\Big|_{\overline{x}}^{+\infty}$$

This means that the corpuscle in "free" motion simultaneously polarizes the "environment" of the wave behind x ≤ ct and in front of x ≥ ct, in such a manner that the resulting force has a symmetrical distribution as compared with the plane that contains the position of the noticeable object x = ct at any time moment t. Under such circumstances, the physical object uniformly moves (the corpuscle is located in the field of the associated wave).

12. The Corpuscle as a Horologe and Its Incorporation in the Associated Wave. Implications

According to de Broglie's theory, the corpuscle must be associated to a horologe having the frequency equal to that of the associated progressive wave. Mathematically we can describe such an oscillator through the differential equation

$$\ddot{q}+\omega^2 q=0 \tag{94}$$

where ω defines the natural frequency of the oscillator as it is dictated by the environment (the wave), and the point above the symbol refers to the differential as compared with time. The most general solution of Equation (94) generally depends not on two arbitrary constants, as it is usually considered, but on three: the initial relevant coordinate, the initial speed and the phase of the harmonic oscillatory within the ensemble that structurally represents the environment (the isolated oscillator is an abstraction). Such a solution gives the relevant co-ordinate

$$q(t)=he^{i(\omega t+\Phi)}+\bar{h}e^{-i(\omega t+\Phi)} \tag{95}$$

where $\bar{h}$ refers to the complex conjugate of h and Φ is an initial phase specific to the individual movement of the oscillator. Such a notation allows us to solve a problem that we could name "the oscillators with the same frequency", such as Planck's resonators' ensemble—the basis of the quantum theory arguments in their old shape. That is, given an ensemble of oscillators having the same frequency in a space region, which is the relation between them?

The mathematical answer to this problem can be obtained if we note that what we want here is to find a mean to pass from a triplet of numbers—the initial conditions—of an oscillator towards the same triplet of another oscillator with the same frequency. This process (passing) implies a simple transitive continuous group with three parameters that can be built using a certain definition of the frequency. We start from the idea that the ratio of two fundamental solutions of Equation (94) is a solution of Schwarts' non linear equation [63-65]

$$\frac{\mathrm{d}}{\mathrm{d}t}\left(\frac{\ddot{\tau}_0}{\dot{\tau}_0}\right)-\frac{1}{2}\left(\frac{\ddot{\tau}_0}{\dot{\tau}_0}\right)^2=2\omega^2,\quad \tau_0(t)\equiv e^{-2i\omega t} \tag{96}$$

This equation proves to be a veritable definition of frequency as a general characteristic of an ensemble of oscillators that can be scanned through a continuous group of three parameters. Indeed Equation (96) is invariant to the change of the dependent variable

$$\tau(t)=\frac{a\tau_0(t)+b}{c\tau_0(t)+\mathrm{d}} \tag{97}$$

which can be verified through direct calculation. Thus, $\tau(t)$ characterizes another oscillator with the same frequency which allows us to say that, starting from a standard oscillator we can scan the whole ensemble of oscillators of the same frequency when we let loose the three ratios a:b:c:d in Equation (97). We can make a more precise correspondence between a homographic change and an oscillator, by means of associating to each oscillator a personal $\tau(t)$ through equation

$$\tau_1(t)=\frac{h+\bar{h}k\tau_0(t)}{1+k\tau_0(t)} \qquad k\equiv e^{-2i\Phi} \tag{98}$$

Let us notice that τ_0, τ_1 can be freely used one instead the other, which leads to the next group of changes for the initial conditions

$$h'\rightarrow\frac{ah+b}{ch+\mathrm{d}} \tag{99a}$$

$$\bar{h}'\rightarrow\frac{a\bar{h}+b}{c\bar{h}+\mathrm{d}} \tag{99b}$$

$$k'\rightarrow k\cdot\frac{c\bar{h}+\mathrm{d}}{ch+\mathrm{d}} \tag{99c}$$

$$a,b,c,d\in R \tag{99d}$$

This is a simple transitive group: one and only one change of the group (the Barbilian group [63,64]) corresponds to a given set of values (a/c, b/c, d/c).

This group admits the 1-differential forms, absolutely invariant through the group [63]

$$\omega_0 = i\left(\frac{\mathrm{d}k}{k} - \frac{\mathrm{d}h + \mathrm{d}\bar{h}}{h - \bar{h}}\right), \omega_1 = \bar{\omega}_2 = \frac{\mathrm{d}h}{k\left(h - \bar{h}\right)} \tag{100}$$

and the 2-differential form

$$\frac{\mathrm{d}s^2}{\alpha^2} = \omega_0{}^2 - 4\omega_1\omega_2 = -\left(\frac{\mathrm{d}k}{k} - \frac{\mathrm{d}h - \mathrm{d}\bar{h}}{h - \bar{h}}\right)^2 + 4\frac{\mathrm{d}h\mathrm{d}\bar{h}}{\left(h - \bar{h}\right)^2},$$

$$\alpha = \text{const.} \tag{101}$$

respectively.

If we restrict the definition of a parallelism of directions in Levi-Civita manner [63]

$$\mathrm{d}\varphi = -\frac{\mathrm{d}u}{v} \tag{102}$$

with

$$h = u + iv,\ \bar{h} = u - iv,\ k = \mathrm{e}^{-i\varphi} \tag{103}$$

Barbilian's group invariates the metrics of Lobacevski's plane [63],

$$\frac{\mathrm{d}s^2}{\alpha^2} = -\frac{\mathrm{d}u^2 + \mathrm{d}v^2}{v^2} \tag{104}$$

Metrics (104) coincides with the differential invariant that is built with the complex scalar field of the speed,

$$\frac{\mathrm{d}s^2}{\alpha^2} \equiv \mathrm{d}\phi d\bar{\phi} = \left(2D\mathrm{d}s - iD\mathrm{d}\ln\rho\right)\left(2D\mathrm{d}s + iD\mathrm{d}\ln\rho\right)$$
$$= 4D^2\left(\mathrm{d}s\right)^2 + D^2\left(\frac{\mathrm{d}\rho}{\rho}\right)^2 \tag{105}$$

which admits the identities

$$\alpha = D, \tag{106a}$$

$$2\mathrm{d}s \equiv \mathrm{d} \quad \Phi = -\frac{\mathrm{d}u}{v}, \tag{106b}$$

$$\mathrm{d}\ln\rho \equiv \mathrm{d}\ln v \tag{106c}$$

Now, through a Matzner-Misner type principle one can obtain Ernst's principle of generating the symmetrical axial metrics [61,62]

$$\delta\int \frac{\nabla h \nabla \bar{h}}{\left(h-\bar{h}\right)^2}\gamma^{1/2}\mathrm{d}^3x = 0 \tag{106d}$$

where g = det g_{ab} with g_{ab} the metrics of the "environment".

Therefore, the incorporation of the corpuscle in the wave, considering that it functions as a horologe with the same frequency as that of the associated progressive wave, implies gravitation through Einstein's vacuum equations (equivalent to Ernst's principle (106d)). On the contrary, when the frequencies do not coincide, there is an induction of Stoler's group from the theory of coherent states (the parameter of the change is the very ratio of frequencies when creation and annihilation operators refer to a harmonic oscillator [63]

Let us note that the homographic changes (99) generalize the result (92). Moreover, if $a,b,c,d \in \mathbf{Z}$ then the Ernst type equations describe super-gravitation N = 1 [65].

13. CONCLUSIONS

Finally we can display the conclusions of this chapter as follows:

- A critical analysis of the hydrodynamic model of Madelung and of the double solution theory of de Broglie's theory of double solution was performed—departing from here, we built a fractal approximation of motion;
- We got the equation of motion of the physical object in the fractal approximation and the Eulerian case was studied;
- The flowing regimes of a rotational fractal fluid were studied;

- We studied the irrotational regime of a fractal fluid and the incorporation of the particle into the associated wave by generating a Schrödinger equation;
- The extended hydrodynamic model of scale relativity was built and the role of the fractal potential in the process of incorporation of the particle into the wave, specified;
- We indicated the mechanisms of wave-particle duality by their in phase coherences;
- We studied the wave-particle duality by stationary flow regimes of a fractal fluid which is coherent in phase, and by non-stationary flow regimes of an incoherent fractal fluid by means of a "polarization" type mechanism;
- Considering the particle as a singularity in the wave, we showed that its incorporation into the associated wave resulted in Einstein's equations in vacuum—contrary, its non-incorporation led to the second quantification;

ACKNOWLEDGEMENTS

This paper was supported by the project POSDRU/88/1.5/ S/47646 of The European Social Fund and the project PERFORM-ERA "Postdoctoral Performance for Integration in the European Research Area" (ID-57649), financed by the European Social Fund and the Romanian Government.

REFERENCES

1. L. Nottale and J. Schneider, "Fractals and Non-Standard Analysis," Journal of Mathematical Physics, Vol. 25, No. 12, 1984, pp. 96-300.
2. L. Nottale, "Fractals and the Quantum Theory of SpaceTime," International Journal of Modern Physics, Vol. A4, No. 50, 1989, pp. 47-117.
3. L. Nottalle, "Fractal Space-Time and Microphysics: Towards a Theory of Scale Relativity," World Scientific, Singapore, 1993.
4. L. Nottale, "Scale Relativity and Fractal Space-Time: Applications to Quantum Physics, Cosmology and Chaotic Systems," Chaos, Solitons & Fractals, Vol. 7, No. 6, 1996, pp. 877-938.

5. L. Nottale, "The Theory of Scale Relativity: Non-Differentiable Geometry, Fractal Space-Time and Quantum Mechanics," Computing Anticipatory Systems: CASYS' 03- 33 Sixth International Conference, Liège, 11-16 August 2003, pp. 68-95.

6. J. Chaline, L. Nottale and P. Grou, "The Trees of Evolution: Universe S," Life Companies Societies, Paris, 2000.

7. J. Chaline, L. Nottale and P. Grou, "Flowers for Scrodinger Sscale Relativity and Its Applications," Editure Ellipses Marketing, Paris, 2009.

8. A. Harabagiu and M. Agop, "Hydrodyamic Model of Scale Relativity Theory," Bulletin of the Polytechnic Institute of Iasi, Vol. LV, No. 3-4, 2005, pp. 77-82.

9. A. Harabagiu, O. Niculescu, M. Colotin, T. D. Bibere, I. Gottlieb and M. Agop, "Particle in a Box by Means of Fractal Hydrodynamic Model," Romanian Reports in Physics, Vol. 61, No. 3, 2009, pp. 395-400.

10. A. Harabagiu, D. Magop and M. Agop, "Fractality and Quantum Mechanics," Ars Longa Publishing House, Iaşi, 2010.

11. M. Agop, L. Chicoş, P. Nica and A. Harabagiu, "Euler's Fluids and Non-Differentiable Space-Time," Far East Journal of Dynamical Systems, Vol. 10, No. 1, 2008, pp. 93-106.

12. M. Agop, A. Harabagiu and P. Nica, "Wave-Particle Duality through a Hydrodynamic Model of the Fractal Space Time Theory," Acta Physica Polonica A, Vol. 113, No. 6, 2008, pp. 1557-1574.

13. S. Ţiţeica, "Quantum Mechanics," Academiei Publishing House, Bucureşti, 1984.

14. B. Felsager, "Geometry, Particles and Fields," Odense University Press, Odense, 1981.

15. A. Peres, "Quantum Theory: Concepts and Methods," Kluwer Academic Publishers, Boston, 1993.

16. J. J. Sakurai and T. S. Fu, "Modern Quantum Mechanics," Addison-Wesley, Boston, 1994.

17. F. Halbwacs, "Theorie Relativiste Des Fluids a Spin," Gauthier-Villars, Paris, 1960.

18. R. Madelung, "A Hydrodynamic Interpretation of Quantum Mechanics," Journal for Physics, Vol. 40, No. 3-4, 1927, pp. 322-326.

19. D. Bohm and B. J. Hiley, "The Undivided Universe: An Ontological Interpretation of Quantum Theory," Routledge and Kegan Paul, London, 1993.

20. D. Dürr, S. Goldstein and N. Zanghi, "Quantum Mechanics, Randomness and Deterministic Reality," Physics Letters A, Vol. 172, No. 1-2, 1992, pp. 6-12.

21. D. Dürr, S. Goldstein and N. Zanchi, "A Global Equilibrium as the Foundation of Quantum Randomness," Foundations of Physics, Vol. 23, 1993, pp. 712-738.

22. D. Z. Albert, "Bohm's Alternative to Quantum Mechanics", Scientific American, Vol. 270, No. 2, 1994, pp. 32- 39.

23. J. S. Bell, "Speakable and Unspeakable in Quantum Mechanics," Cambridge University Press, Cambridge, 1987.

24. K. Berndl, D. Dürr, S. Goldstein, G. Peruzzi and N. Zanchi, "Existance of Trajectories for Bohmian Mechanics," International Journal of Theoretical Physics, Vol. 32, No. 12, 1993, pp. 2245-2251.

25. K. Berndl, D. Dürr, S. Goldstein, N. Zanchi, "Selfadjointness and the Existance of Deterministic Trajectories in Quantum Theory," In: M. Fannes, C. Maes and A. Verbeure, Eds., On Three Levels: Micro-, Meso-, and Macroscopic Approches in Physics, (NATO ASI Series B: Physics), Plenum, New York, 1994, pp. 429-434.

26. P. R. Holland, "The Quantum Theory of Motion," Cambridge University Press, Cambridge, 1993.

27. L. de Broglie, "Causal Interpretation Attempt and NonLinear Wave Mechanics: The Theory of Double Solution," Gauthier-Villars, Paris, 1956.

28. L. de Broglie, "The Measure Theory in Wave Mechanics," Gauthier-Villars, Paris, 1957.

29. L. de Broglie, "Wave Mechanics Interpretation," Journal of Physics Radium, Vol. 20, No. 12, 1959, pp. 963-979.

30. L. de Broglie, "Critical Study of the Current Foundations Interpretation of Wave Mechanics," Gauthier-Villars, Paris, 1963.

31. L. de Broglie, "Thermodynamics of Isolated Particle (Hidden Thermodynamics of Particles)," Gauthier-Villars, Paris, 1964.

32. L. de Broglie, "Certainity and Uncertainity in Science," Politică Publishing House, Bucureşti, 1980.

33. D. Bohm, "A Suggested Interpretation of Quantum Theory in Terms of 'Hidden' Variables I," Physical Review, Vol. 85, No. 2, 1952, pp. 166-179.

34. D. Bohm, "A Suggested Interpretation of Quantum Theory in Terms of 'Hidden Variables': Part II," Physical Review, Vol. 85, No. 2, 1952, pp. 180-193.

35. D. Bohm, "Proof that Probability Density Approaches $|\psi|$ in Causal Interpretation of Quantum Theory," Physical Review, Vol. 89, No. 2, 1953, pp. 458-466.

36. L. Budei, "Fractal Models: Applications in Enviroment Arhitecture," Publishing House University, Gh. Asachi, Iaşi, 2000.

37. M. Barnsley, "Fractals Everywhere: The First Course in Deterministic Fractal Geometry," Academic Press, Boston, 1988.

38. A. Le Mehante, "Les Geometries Fractales," Hermes, Paris, 1990.

39. A. Heck and J. M. Perdang, "Applying Fractals in Astronomy," Springer Verlag, Berlin, 1991.

40. J. Feder and A. Aharony, "Fractals in Physics," NorthHolland, Amsterdam, 1990.

41. P. Berge, Y. Pomeau and C. Vidal, "The Order in Chaos," Wiley, New York, 1984.

42. J. F. Gouyet, "Physical Structures and Fractals," Masson, Paris, 1992.

43. M. S. El Naschie, O. E. Rössler and I. Prigogine, "Quantum Mechanics, Diffusion and Chaotic Fractals", Elsevier, Oxford, 1995.

44. P. Weibel, G. Ord and O. E. Rössler, "Space Time Physics and Fractality: Festschroft in Honer of Mohamad El Naschie Vienna," Springer, New York, 2005.

45. E. Nelson, "Quantum Fluctuations," Princeton University Press, Princeton, 1985.

46. M. Agop, M. Colotin and V. Păun, "Haoticity, Fractality and Fields: Elements of Fractal Theory," Publishing House ArsLonga, Iaşi, 2009, pp. 12-46.

47. V. Chioroiu, L. Munteanu, P. Ştiucă and S. Donescu, "Introduction in Nanomechanics," Publishing House Academiei Române, Bucureşti, 2005.

48. D. K. Ferry and S. M. Goodnick, "Transport in Nanostructures," Cambridge University Press, Cambridge, 2001.

49. Y. Imry, "Introduction to Mesoscopic Physics," Oxford University Press, Oxford, 2002.

50. D. Benoit, "Physics of Semiconductor Microcavities from Fundamentals to Nano-Scale Decretes," Wiley-VCH Verlag GmbH, Weinheim, 2007.

51. C. Ciuti and I. Camsotto, "Quantum Fluids Effects and Parametric Instabilities in Microcavities," Physica Status Solidi B, Vol. 242, No. 11, 2005, p. 2224.

52. C. P. Poole, K. A. Farach and R. Creswick, "Super-Conductivity," Academic Press, San Diego, 1995.

53. G. Grössing, "Diffusion Waves in Sub-Quantum Thermodynamics: Resolution of Einstein's 'Particle-in-a-Box' Objection," 2008.

54. A. Mandelis, L. Nicolaides and Y. Chen, "Structure and the Reflectionless/Refractionless Nature of Parabolic Diffusion-Wave Fields, " Physical Review Letters, Vol. 87, No. 2, 2001, 4 pages.

55. A. Mandelis, "Diffusion Waves and Their Uses," Physics Today, Vol. 53, No. 8, 2000, p. 29.

56. S. Popescu, "Actual Issues in Self-Structuring Systems," Publishing House Tehnopress, Iasi, 2004.

57. E. R. Bittner, "Quantum Tunneling Dynamics Using Hydrodynamic Trajectories," Journal of Chemical Physics, Vol. 112, No. 22, 2000, p. 9703.

58. D. Ruelle and F. Takens, "On the Nature of Turbulence," Communications in Mathematical Physics, Vol. 20, No. 23, 1971, pp. 167-192,

59. D. Ruelle, "Strange Attractors," The Mathematical Intelligencer, Vol. 2, No. 3, 1979, pp. 126-137.

60. C. V. Munceleanu, D. Magop, C. Marin and M. Agop, "Fractal Models in Polymer Physics," Ars Longa Publishing House, Iasi, 2010.

61. F. J. Ernst, "New Formulation of the Axially Symemetric Gravitational Fielf Problem I," Physical Reviews, Vol. 167, No. 5, 1968, pp. 1175-1178.

62. F. J. Ernst, "Exterior Algebraic Derivation of Einstein Field Equation Employing a Generalized Basis," Journal of Mathematical Physics, Vol. 12, No. 11, 1971, pp. 2395-2398.

63. M. Agop and N. Mazilu, "Fundaments of Modern Physics," Junimea Publishing House, Iasi, 1989.

64. M. Agop and N. Mazilu, "The Crossroads of the Theories between Newton and Einstein Barbilian's Univers," Ars Longa, Iaşi, 2010.

65. M. B. Green, J. H. Schwarz and E. Witten, "Superstring Theory," Cambridge University Press, Cambridge, 1998.

5

CHAPTER

DISCRETE SYMMETRY IN RELATIVISTIC QUANTUM MECHANICS

Guang-jiong Ni[1,2], Suqing Chen[2], Jianjun Xu[2]

[1]*Department of Physics, Portland State University, Portland, USA*

[2]*Department of Physics, Fudan University, Shanghai, China*

ABSTRACT

EPR experiment on $K^0-\overline{K}^0$ system in 1998 [1] strongly hints that one should use operators $\hat{E}_c=-i\hbar\frac{\partial}{\partial t}$ and $\hat{\boldsymbol{p}}_c=i\hbar\nabla$ for the wavefunction (WF) of antiparticle. Further analysis on Klein-Gordon (KG) equation reveals that there is a discrete symmetry hiding in relativistic quantum mechanics (RQM) that $\mathcal{PT}=C$. Here $\mathcal{PT}$ means the (newly defined) combined space-time inversion (with $x\rightarrow -x, t\rightarrow -t$), while C the transformation of WF ψ between particle and its antiparticle whose definition is just residing in the above symmetry. After combining with FeshbachVillars (FV) dissociation of KG equation $(\psi=\phi+\chi)$ [2], this discrete symmetry can be rigorously reformulated by the invariance of coupling equation of ϕ and

χ under either the combined space-time inversion $\mathcal{PT}$ or the mass inversion $(m \to -m)$, which makes the KG equation a self-consistent theory. Dirac equation is also discussed accordingly. Various applications of this discrete symmetry are discussed, including the prediction of antigravity between matter and antimatter as well as the reason why we believe neutrinos are likely the tachyons.

Keywords: CPT Invariance; Antiparticle; Quantum Mechanics; Quantum Field Theory

1. INTRODUCTION

In 1956-1957, the historical discovery of the parity violation [3-6] reveals that both P and C symmetries are violated to maximum in weak interactions. Then in 1964- 1970, both CP and T are experimentally verified to be violated in some cases (though to a tiny degree) [7,8] whereas the product symmetry CPT holds intact to this day [9]. The CPT invariance in quantum field theory (QFT) was first proved by Lüders and Pauli in 1954-1957 [10-12] via the introduction of the "strong reflection" for proving the CPT theorem. In 1965, Lee and Wu proposed that the definition of particle $|a\rangle$ versus its antiparticle $|\bar{a}\rangle$ should be [13]

$$|\bar{a}\rangle = \mathrm{CPT}|a\rangle \tag{1.1}$$

Regrettably, the counterpart of "strong reflection" at the level of RQM went nearly unnoticed in the past decades. In this paper, we are going to study the RQM thoroughly. Not only a discrete symmetry $\mathcal{PT} = \mathcal{C}$ is found in RQM as the counterpart of "strong reflection" in QFT, it is also evolved into the invariance of space-time inversion $(x \to -x, t \to -t)$ or mass inversion $(m \to -m)$, showing that a WF in RQM is always composed of two parts in confrontation inside a particle and then RQM becomes a self-consistent theory. Furthermore, this symmetry can serve as a "theoretical tool" in searching for new applications in today's physics.

The organization of this paper is as follows: In section II, the EPR paradox [14] is discussed together with the $K^0 - \bar{K}^0$ correlation experimental data

[1], yielding a strong hint that the energy-momentum operators for antiparticle's WF should be $\hat{E}_c = -i\hbar\frac{\partial}{\partial t}$ and $\hat{\boldsymbol{p}}_c = i\hbar\nabla$ respectively. Section III is focused on a discrete symmetry $\mathcal{PT} = \mathcal{C}$, here $\mathcal{PT}$ means the (newly defined) combined space-time inversion (with $\boldsymbol{x} \to -\boldsymbol{x}, t \to -t$), while $\mathcal{C}$ the transformation of WFs between particle and antiparticle, whose definition is just residing in the symmetry. Then after combining with FV dissociation of KG equation [2] in which the WF ψ is composed of two fields: $\psi = \phi + \chi$, the above symmetry can be realized in terms of ϕ and χ rigorously via the invariance of their coupling equation either under the spacetime inversion or a mass inversion $(m \to -m)$ In this way, the probability density is ensured to be positive definite for WFs of either particle or antiparticle. Section IV ascribes various phenomena in the theory of special relativity (SR) to the effects of enhancement of the hidden χ field in a moving particle. In Section V, Dirac equation is discussed accordingly with the importance of helicity being stressed. Section VI contains a brief discussion on the QFT. Sections VII, VIII and IX are devoting to seek for possible applications of the above symmetry in today's physical problems: Why a parity violation phenomenon was overlooked since 1956-1957? Why we believe neutrinos are likely the tachyons? And the prediction of antigravity between matter and antimatter. The last Section X contains a summary. In the Appendix, the Klein paradox is solved for both KG equation and Dirac equation without resorting to the "hole theory".

2. WHAT THE $K^0\bar{K}^0$ CORRELATION EXPERIMENTAL DATA ARE TELLING?

To our knowledge, beginning from Bohm and Bell [15,16], physicists gradually turned their research of EPR paradox [14] onto the entangled state composed of electrons, especially photons with spin and achieved fruitful results. However, as pointed out by Guan (1935-2007), EPR's paper [14] is focused on two spinless particles and Guan found that there is a commutation relation hiding in such a system as follows [17]:

Consider two particles in one dimensional space with positions $x_i\,(i = 1,2)$ and momentum operators

$\hat{p}_i = -i\hbar \frac{\partial}{\partial x_i}$. Then a commutation relation arises as

$$[x_1 - x_2, \hat{p}_1 + \hat{p}_2] = 0 \quad (2.1)$$

According to QM's principle, there may be a kind of common eigenstate having eigenvalues of these two commutative (i.e., compatible)observables like:

$$p_1 + p_2 = 0, (p_2 = -p_1) \text{ and } (x_1 - x_2) = D \quad (2.2)$$

with D being their distance. The existence of such kind of eigenstate described by Equation (2.2) puzzled Guan, he asked: "How can such kind of quantum state be realized?" A discussion between Guan and one of present authors (Ni) in 1998 led to a paper [18].

Here we are going to discuss further, showing that the correlation experiment on a $K^0\overline{K}^0$ system (which just realized an entangled state composed of two spinless particles) in 1998 by CPLEAR collaboration [1] actually revealed some important features of QM and then answered the puzzle raised by EPR in a surprising way. First, besides Equation (1), let us consider another three commutation relations simultaneously:

$$[t_1 + t_2, \hat{E}_1 - \hat{E}_2] = 0 \quad (2.3)$$

$$[x_1 + x_2, \hat{p}_1 - \hat{p}_2] = 0 \quad (2.4)$$

$$[t_1 - t_2, \hat{E}_1 + \hat{E}_2] = 0 \quad (2.5)$$

($E_i = i\hbar \frac{\partial}{\partial t_i}$ with t_i being the time during which the i-th particle is detected). In accordance with Ref. [1], we also focus on back-to-back events. The evolution of $K^0\overline{K}^0$'s wavefunction (WF) will be considered in three inertial frames: The center-of-mass system S is at rest in laboratory with its origin x = 0 located at the apparatus' center, where the antiprotons' beam is stopped inside a hydrogen gas target to create $K^0\overline{K}^0$ pairs by $p\overline{p}$ annihilation. The $K^0\overline{K}^0$ pairs are detected by a cylindrical tracking detector located inside a solenoid providing a magnetic field parallel to the antiprotons' beam. For back-to-back events, the space-time coordinates in Equations (1)-(5) refer to particles moving to the right $(x_1 > 0)$ and left $(x_2 < 0)$ respectively. Second, we take an inertial system S' with its origin located at particle 1 (i.e., $x_1' = 0$). S' is moving in a uniform velocity v with respect to S. (For Kaon's momentum of $800\ \mathrm{MeV/c}, \beta = v/c = 0.849$). Another S'' system is chosen with its origin located at particle $2(x_2'' = 0)$. S'' is moving in a velocity $(-v)$ with respect to S. Thus we have Lorentz transformation among the space-time coordinates being

$$\begin{cases} x' = \dfrac{x - vt}{\sqrt{1-\beta^2}}, \\ t' = \dfrac{t - vx/c^2}{\sqrt{1-\beta^2}}, \end{cases} \begin{cases} x'' = \dfrac{x + vt}{\sqrt{1-\beta^2}}, \\ t'' = \dfrac{t + vx/c^2}{\sqrt{1-\beta^2}}, \end{cases} \tag{2.6}$$

Here t_1' and t_2'' correspond to the proper time t_a and t_b in Ref.[1] respectively. The common time origin $t = t' = t'' = 0$ is adopted.

A $K^0\bar{K}^0$ pair, created in a $J^{PC} = 1^{--}$ antisymmetric state, can be described by a two-body WF depending on time as ([1], see also [19,20])

$$\begin{aligned} &|\Psi(0,0)\rangle^{(\text{antisym})} \\ &= \frac{1}{\sqrt{2}}\left[|K^0(0)\rangle_a |\bar{K}^0(0)\rangle_b - |\bar{K}^0(0)\rangle_a |K^0(0)\rangle_b\right] \\ &|\Psi(t_a,t_b)\rangle^{(\text{antisym})} \\ &= \frac{1}{\sqrt{2}}\Big[|K_S(0)\rangle_a |K_L(0)\rangle_b\, \mathrm{e}^{-i(\alpha_S t_a + \alpha_L t_b)} \\ &\qquad - |K_L(0)\rangle_a |K_S(0)\rangle_b\, \mathrm{e}^{-i(\alpha_L t_a + \alpha_S t_b)}\Big] \end{aligned} \tag{2.7}$$

with

$$|K_S\rangle = \frac{1}{\sqrt{2}}\left[|K^0\rangle - |\bar{K}^0\rangle\right], |K_L\rangle = \frac{1}{\sqrt{2}}\left[|K^0\rangle + |\bar{K}^0\rangle\right] \tag{2.8}$$

where the CP violation has been neglected and

$\alpha_{S,L} = m_{S,L} - \mathrm{i}\gamma_{S,L}/2$, $m_{S,L}$ and $\gamma_{S,L}$ being the $K_{S,L}$ masses and decay widths, respectively. From Equation (7), the intensities of events with like-strangeness (K^0K^0 or $\bar{K}^0\bar{K}^0$) and unlike-strangeness ($K^0\bar{K}^0$ or $\bar{K}^0K^0$) can be evaluated as

$$I_{\text{like}}^{(\text{antisy})}(t_a, t_b) = \frac{1}{8}\mathrm{e}^{-2\gamma\tilde{t}} \cdot \left\{\mathrm{e}^{-\gamma_S|t_a - t_b|} + \mathrm{e}^{-\gamma_L|t_a - t_b|} - 2\mathrm{e}^{-\gamma|t_a - t_b|}\cos\left[\Delta m(t_a - t_b)\right]\right\} \quad (2.9)$$

$$I_{\text{unlike}}^{(\text{antisy})}(t_a, t_b) = \frac{1}{8}\mathrm{e}^{-2\gamma\tilde{t}} \cdot \left\{\mathrm{e}^{-\gamma_S|t_a - t_b|} + \mathrm{e}^{-\gamma_L|t_a - t_b|} + 2\mathrm{e}^{-\gamma|t_a - t_b|}\cos\left[\Delta m(t_a - t_b)\right]\right\} \quad (2.10)$$

where $\Delta m = m_L - m_S, \gamma = (\gamma_S + \gamma_L)/2$ and $\tilde{t} = t_a\,(\text{for } t_a < t_b)$ or $\tilde{t} = t_b\,(\text{for } t_a > t_b)$.

Similarly, for $K^0\bar{K}^0$ created in a $J^{PC} = 0^{++}$ or 2^{++} symmetric state as:

$$|\Psi(0,0)\rangle^{(sym)} = \frac{1}{\sqrt{2}}\left[|K^0(0)\rangle_a|\bar{K}^0(0)\rangle_b + |\bar{K}^0(0)\rangle_a|K^0(0)\rangle_b\right]$$

$$|\Psi(t_a,t_b)\rangle^{(sym)} = \frac{1}{\sqrt{2}}\left[|K_L(0)\rangle_a|K_L(0)\rangle_b\,\mathrm{e}^{-\mathrm{i}(\alpha_L t_a + \alpha_L t_b)} - |K_S(0)\rangle_a|K_S(0)\rangle_b\,\mathrm{e}^{-\mathrm{i}(\alpha_S t_a + \alpha_S t_b)}\right] \quad (2.11)$$

the predicted intensities read

$$I_{\text{like}}^{(sym)}(t_a,t_b)$$
$$=\frac{1}{8}\left\{e^{-\gamma_S(t_a+t_b)}+e^{-\gamma_L(t_a+t_b)}-2e^{-\gamma(t_a+t_b)}\cos\left[\Delta m(t_a+t_b)\right]\right\}$$
$$I_{\text{unlike}}^{(sym)}(t_a,t_b)$$
$$=\frac{1}{8}\left\{e^{-\gamma_S(t_a+t_b)}+e^{-\gamma_L(t_a+t_b)}+2e^{-\gamma(t_a+t_b)}\cos\left[\Delta m(t_a+t_b)\right]\right\} \quad (2.12)$$

The experiment [1] reveals that the $K^0\overline{K}^0$ pairs are mainly created in the antisymmetric state shown by Equations (2.9) and (2.10) while the contribution in a symmetric state shown by Equations (2.11) and (2.12) accounts for 7.4%.

What we learn from Ref. [1] in combination with Equations (2.1)-(2.5) are as follows:

(a) Because only back-to-back events are involved in the S system, we denote three commutative operators as: the "distance" operator

$\hat{D}=x_1-x_2=v(t_1+t_2)$, $\hat{A}=\hat{p}_1+\hat{p}_2$ and $\hat{B}=\hat{E}_1-\hat{E}_2$, Equations (2.1) and (2.3) read

$$\left[\hat{D},\hat{A}\right]=0,\left[\hat{D},\hat{B}\right]=0,\left[\hat{A},\hat{B}\right]=0 \quad (2.13)$$

So they may have a kind of common eigenstate during the measurement composed of K^0K^0 and projected from the symmetric state shown by Equation (11). It is assigned by a continuous eigenvalue $D_j=v(t_1+t_2)$ (with continuous index j) of operator $\hat{D}$ acting on the WF, $\Psi_{K^0K^0}^{sym}(x_1,t_1;x_2,t_2)$, as[1]

$$\hat{D}\Psi_{K^0K^0}^{sym}(x_1,t_1;x_2,t_2)$$
$$=D_j\Psi_{K^0K^0}^{sym}(x_1,t_1;x_2,t_2)$$
$$=v(t_1+t_2)\Psi_{K^0K^0}^{sym}(x_1,t_1;x_2,t_2) \quad (2.14a)$$

$$\hat{A}\Psi^{sym}_{K^0K^0}(x_1,t_1;x_2,t_2)$$
$$=A_j^{\text{like}}\Psi^{sym}_{K^0K^0}(x_1,t_1;x_2,t_2)$$
$$=(p_1+p_2)\Psi^{sym}_{K^0K^0}(x_1,t_1;x_2,t_2) \tag{2.15}$$

$$\hat{B}\Psi^{sym}_{K^0K^0}(x_1,t_1;x_2,t_2)$$
$$=B_j^{\text{like}}\Psi^{sym}_{K^0K^0}(x_1,t_1;x_2,t_2)$$
$$=(E_1-E_2)\Psi^{sym}_{K^0K^0}(x_1,t_1;x_2,t_2) \tag{2.16}$$

where the lowest eigenvalue of $\hat{A}$ is

$A_j^{\text{like}}=p_1+p_2=0,(p_2=-p_1)$, and that of $\hat{B}$ is $B_j^{\text{like}}=E_1-E_2=0,(E_2=E_1)$ respectively. These eigenstates of like-strangeness events predicted by Equation (11) are really observed in the experiment [1] (these eigenstates of K^0K^0 were overlooked in the Ref. [18]).

(b) The more interesting case occurs for $K^0\bar{K}^0$ pair created in the antisymmetric state with intensity given by Equation (10) being a function of (t_a-t_b) (not (t_a+t_b) as shown by Equation (12) for symmetric states)

which is proportional to (t_1-t_2) in the S system. In the EPR limit $t_1=t_2$, $K^0\bar{K}^0$ events dominate whereas likestrangeness events are strongly suppressed as shown by Equation (9) (see Figure 1 in [1]). So the experimental facts remind us of the possibility that $K^0\bar{K}^0$ events may be related to common lowest (zero) eigenvalues of some commutative operators (just like what happened in Equations (15) and (16) for operators $\hat{A}$ and $\hat{B}$ (which are applied to symmetric states (due to

$\hat{D}=x_1-x_2=v(t_1+t_2)$) but are not suitable for antisymmetric states), there are another three operators shown by Equations (4) and (5) being: the operator of "flight-path difference" $\hat{F}=x_1+x_2=v(t_1-t_2),\hat{M}=\hat{p}_1-\hat{p}_2$ and $\hat{G}=\hat{E}_1+\hat{E}_2$ with commutation relations as:

$$\left[\hat{F},\hat{M}\right]=0,\left[\hat{F},\hat{G}\right]=0,\left[\hat{M},\hat{G}\right]=0 \tag{2.17}$$

which are just suitable for antisymmetric states. For $K^0\bar{K}^0$ back-to-back events, assume that one of two particles, say 2, is an antiparticle with its momentum and energy operators being

$$\hat{p}_x^c = \mathrm{i}\hbar\frac{\partial}{\partial x}, \hat{E}^c = -\mathrm{i}\hbar\frac{\partial}{\partial t} \tag{2.18}$$

(the superscript c means "antiparticle") versus that for particle being

$$\hat{p}_x = -\mathrm{i}\hbar\frac{\partial}{\partial x}, \hat{E} = \mathrm{i}\hbar\frac{\partial}{\partial t} \tag{2.19}$$

For instance, a freely moving particle's WF reads[2]:

$$\psi(x,t) \sim \exp\left[\frac{\mathrm{i}}{\hbar}(px - Et)\right] \tag{2.20}$$

whereas

$$\psi_c(x,t) \sim \exp\left[-\frac{\mathrm{i}}{\hbar}(p_c x - E_c t)\right] \tag{2.21}$$

for its antiparticle with $p_c = p$ and $E_c(>0)$ being momentum and energy of the antiparticle in accordance with Equation (2.18). If using Equations (2.18)-(2.21), we find

$$\begin{aligned}&\hat{F}\Psi_{K^0\bar{K}^0}^{\text{antisym}}(x_1,t_1;x_2,t_2)\\&= F_k^{\text{unlike}}\Psi_{K^0\bar{K}^0}^{\text{antisym}}(x_1,t_1;x_2,t_2)\\&= v(t_1-t_2)\Psi_{K^0\bar{K}^0}^{\text{antisym}}(x_1,t_1;x_2,t_2)\end{aligned} \tag{2.22}$$

with continuous index k referring to continuous eigenvalues $F_k = v(t_1 - t_2)$. Here, the WF in space-time of this system during measurement reads approximately:

$$\Psi^{\text{antisym}}_{K^0\overline{K}^0}\left(x_1,t_1;x_2,t_2\right)\sim \mathrm{e}^{\mathrm{i}(p_1x_1-E_1t_1)}\mathrm{e}^{-\mathrm{i}\left(p_2^cx_2-E_2^ct_2\right)} \tag{2.23}$$

with antiparticle 2 moving opposite to particle 1 and $p_2^c=-p_1$.

Now we use $\hat{M}\left(=\hat{p}_1-\hat{p}_2\right)=\hat{p}_1+\hat{p}_2^c$ on $K^0\overline{K}^0$ system, yielding

$$\begin{aligned}&\hat{M}\Psi^{\text{antisym}}_{K^0\overline{K}^0}\left(x_1,t_1;x_2,t_2\right)\\&=M_k^{\text{unlike}}\Psi^{\text{antisym}}_{K^0\overline{K}^0}\left(x_1,t_1;x_2,t_2\right)\\&=\left(p_1+p_2^c\right)\Psi^{\text{antisym}}_{K^0\overline{K}^0}\left(x_1,t_1;x_2,t_2\right)\end{aligned} \tag{24}$$

Similarly, we have $\hat{G}\left(=\hat{E}_1+\hat{E}_2\right)=\hat{E}_1-\hat{E}_2^c$ and find

$$\begin{aligned}&\hat{G}\Psi^{\text{antisym}}_{K^0\overline{K}^0}\left(x_1,t_1;x_2,t_2\right)\\&=G_k^{\text{unlike}}\Psi^{\text{antisym}}_{K^0\overline{K}^0}\left(x_1,t_1;x_2,t_2\right)\\&=\left(E_1-E_2^c\right)\Psi^{\text{antisym}}_{K^0\overline{K}^0}\left(x_1,t_1;x_2,t_2\right)\end{aligned} \tag{25}$$

Hence we see that once Equations (2.18) and (2.21) are accepted, the WFs $\Psi^{\text{antisym}}_{K^0\overline{K}^0}\left(x_1,t_1;x_2,t_2\right)$ show up in experiments as the only WFs with strongest intensity at the EPR limit $\left(t_1=t_2\right)$ corresponding to their three eigenvalues being all zero: $F_k=M_k^{\text{unlike}}=G_k^{\text{unlike}}=0$ and they won't change even when accelerator's energies are going up.

If using Equation (2.18), the eigenvalues of $\hat{A}$ and $\hat{B}$ for the WF $\Psi^{\text{antisym}}_{K^0\overline{K}^0}\left(x_1,t_1;x_2,t_2\right)$ are $A_j^{\text{unlike}}=p_1-p_2^c=2p_1$ and $B_j^{\text{unlike}}=E_1+E_2^c=2E_1$ respectively, while that of $\hat{M}$ and $\hat{G}$ for the WF $\Psi^{\text{antisym}}_{K^0K^0}\left(x_1,t_1;x_2,t_2\right)$ are $M_k^{\text{like}}=p_1-p_2=2p_1$ and

$G_k^{\text{like}} = E_1 + E_2 = 2E_1$, respectively, those eigenvalues are much higher than zero and going up with the accelerator's energy.

Something is very interesting here: If we deny Equation (2.18) but insist on unified operators $\hat{p}$ and $\hat{E}$ for both particle and antiparticle, there would be no difference in eigenvalues between like-strangeness events and unlike-strangeness ones. For example, the M_k^{unlike} and G_k^{unlike} would be $2p_1$ and $2E_1$ too (instead of "0" as in Equations (2.24) and (2.25)). This would mean that three commutative operators $\hat{F}, \hat{M}$ and $\hat{G}$ are not enough to distinguish the WF $\Psi_{K^0\bar{K}^0}^{\text{antisym}}(x_1, t_1; x_2, t_2)$ from the WF $\Psi_{K^0K^0}^{\text{antisym}}(x_1, t_1; x_2, t_2)$ even they behave so differently as shown by Equations (9) and (10)), especially at the EPR limit $(t_1 = t_2)$.

Equation (2.18) together with the identification of WF $\Psi_{K^0\bar{K}^0}^{\text{antisym}}(x_1, t_1; x_2, t_2)$ by three zero eigenvalues implies that the difference of a particle from its antiparticle is not something hiding in the "intrinsic space" like opposite charge (for electron and positron) or opposite strangeness (for K^0 and $\bar{K}^0$) but can be displayed in their WFs evolving in space-time at the level of QM.

In summary, instead of one set of WF with its operators (Equations (2.19) and (2.20)), two sets of WFs with operators separately (shown as Equations (2.18)- (2.21)) are strongly supported by the original EPR paradox and its "solution" provided by the $K^0 - \bar{K}^0$ experiment.

To our knowledge, Equation (2.18) can be found at a page note of a paper by Konopinski and Mahmaud in 1953 [21], also appears in Refs. [18,22-28].

3. How to Make Klein-Gordon Equation a Self-Consistent Theory in RQM? A Discrete Symmetry

3.1. The Negative Energy Solution and the WF of Antiparticle

Let us begin with the energy conservation law for a particle in classical mechanics:

$$E = \frac{1}{2m}\boldsymbol{p}^2 + V(x) \tag{3.1}$$

Consider the rule promoting observables into operators:

$$E \to \hat{E} = \mathrm{i}\hbar\frac{\partial}{\partial t}, \boldsymbol{p} \to \hat{\boldsymbol{p}} = -\mathrm{i}\hbar\nabla \tag{3.2}$$

and let Equation (3.1) act on a wavefunction (WF) $\psi(\boldsymbol{x},t)$, the Schrödinger equation

$$\mathrm{i}\hbar\frac{\partial}{\partial t}\psi(\boldsymbol{x},t) = -\frac{\hbar^2}{2m}\nabla^2\psi(\boldsymbol{x},t) + V(\boldsymbol{x})\psi(\boldsymbol{x},t) \tag{3.3}$$

follows immediately. In mid 1920's, considering the kinematical relation for a particle in the theory of special relativity (SR):

$$(E - V)^2 = c^2\boldsymbol{p}^2 + m^2c^4 \tag{3.4}$$

and using Equation (3.2) again, the Klein-Gordon (KG) equation was established as:

$$\left(\mathrm{i}\hbar\frac{\partial}{\partial t} - V\right)^2 \psi(x,t) = -c^2\hbar^2\nabla^2\psi(x,t) + m^2c^4\psi(x,t) \tag{3.5}$$

For a free KG particle, its plane-wave solution reads:

$$\psi(\boldsymbol{x},t) \sim \exp\left[\frac{\mathrm{i}}{\hbar}(\boldsymbol{p}\cdot\boldsymbol{x} - Et)\right] \tag{3.6}$$

However, two difficulties arose:

(a) The energy E in Equation (6) has two eigenvalues:

$$E = \pm\sqrt{c^2\boldsymbol{p}^2 + m^2c^4} \tag{3.7}$$

In general, $V \neq 0$, the WFs of KG particle's energy eigenstates can always be divided into two parts:

$$\psi \sim \exp\left(-\frac{i}{\hbar}Et\right), E > 0 \tag{3.8}$$

$$\psi \sim \exp\left(-\frac{i}{\hbar}Et\right), E < 0 \tag{3.9}$$

where only the original operators Equation (3.2) are used. But what the "negative energy" means?

(b) The continuity equation is derived from Equation (5) as

$$\frac{\partial \rho}{\partial t} + \nabla \cdot \boldsymbol{j} = 0 \tag{3.10}$$

where

$$\rho = \frac{i\hbar}{2mc^2}\left(\psi^* \frac{\partial}{\partial t}\psi - \psi \frac{\partial}{\partial t}\psi^*\right) - \frac{1}{mc^2}V\psi^*\psi \tag{3.11}$$

and

$$\boldsymbol{j} = \frac{i\hbar}{2m}\left(\psi\nabla\psi^* - \psi^*\nabla\psi\right) \tag{3.12}$$

are the "probability density" and "probability current density" respectively. While the latter is the same as that derived from Equation (3.3), Equation (3.11) seems not positive definite and dramatically different from $\rho = \psi^*\psi$ in Equation (3.3). Why?

In hindsight, for a linear equation in RQM, either KG or Dirac equation, the emergence of WFs with both positive and negative energy (E) is inevitable and natural. From mathematical point of view, the set of WFs cannot be complete if without taking the negative energy solutions into account. And physicists believe that these negative-energy solutions might be relevant to antiparticles. However, we physicists admit that both a rest particle's energy $E = mc^2$ and a rest antiparticle's energy $E_c = m_c c^2 = mc^2$ are positive, as verified by numerous experiments like that of pair-creation process $\gamma \to e^+ + e^-$. The above contradiction constructs socalled "negative-energy paradox" in RQM. For Dirac particle, majority (not all) of physicists accept the "hole theory" to explain the "paradox". But for KG particle, no such kind of "hole theory" can be acceptable. It was this "negative-energy paradox" as well as the four "commutation relations", Equations (2.1)-(2.5), hidden in the two-particle system discussed by EPR [14] gradually prompted us to realize that the root cause of difficulty in RQM lies in an a priori notion—only one kind of WF with one set of operators (like Equation (3.2)) can be acceptable in QM, either for NRQM or RQM.

Once getting rid of the constraint in the above notion and introducing two sets of WFs and operators for particle and antiparticle respectively, we can identify the negative energy solution, Equation (3.9), with the antiparticle's WF directly

$$\psi_c \sim \exp\left(\frac{i}{\hbar}E_c t\right), E_c > 0 \tag{3.13}$$

which implies an antiparticle with positive energy E_c by using Equation (2.18). This claim will be proved rigorously in the next subsection.

One may ask: When you assume the negative energy solution being the WF of antiparticle, how about the difficulty of negative probability density? Below we will see how to solve these two difficulties simultaneously and make KG equation a self-consistent theory at the level of RQM.

3.2. The Proof of a Discrete Symmetry $\mathcal{PT} = \mathcal{C}$ for KG Particle

Let us introduce an operator of (newly defined) combined space-time inversion $\mathcal{PT}$ for KG equation. It should change the space-time coordinates as

$$\boldsymbol{x} \to -\boldsymbol{x}, t \to -t \tag{3.14}$$

then accordingly

$$\begin{aligned} \hat{\boldsymbol{p}} &= -i\hbar\nabla \to \mathcal{PT}\hat{\boldsymbol{p}}(\mathcal{PT})^{-1} = \hat{\boldsymbol{p}}_c = i\hbar\nabla, \\ \hat{E} &= i\hbar\frac{\partial}{\partial t} \to \mathcal{PT}\hat{E}(\mathcal{PT})^{-1} = \hat{E}_c = -i\hbar\frac{\partial}{\partial t} \end{aligned} \tag{3.15}$$

Because the antiparticle has opposite charge $(-q)$ versus q for particle, so

$$V(\boldsymbol{x},t) \to \mathcal{PT}V(\boldsymbol{x},t)(\mathcal{PT})^{-1}$$
$$\equiv V_c(\boldsymbol{x},t) = -V(\boldsymbol{x},t) \tag{3.16}$$

When performing $\mathcal{PT}$ inversion on KG equation, Equation (3.5), from left to right, we meet eventually the WF and define the antiparticle's WF as

$$\mathcal{PT}\psi(\boldsymbol{x},t) \equiv \mathcal{C}\psi(\boldsymbol{x},t) = \psi_c(\boldsymbol{x},t) \tag{3.17}$$

Thus KG particle's equation, Equation (3.5), is transformed into $(\hbar = 1)$

$$\left(\hat{E}_c - V_c\right)^2 \psi_c(\boldsymbol{x},t)$$
$$= -c^2\nabla^2\psi_c(\boldsymbol{x},t) + m^2\psi_c(\boldsymbol{x},t) \tag{3.18}$$

or

$$\left(\mathrm{i}\frac{\partial}{\partial t} - V\right)^2 \psi_c(\boldsymbol{x},t)$$
$$= -c^2\nabla^2\psi_c(\boldsymbol{x},t) + m^2\psi_c(\boldsymbol{x},t) \tag{3.19}$$

which is formally the same as Equation (3.5) though we should use $\hat{\boldsymbol{p}}_c, \hat{E}_c$ for $\psi_c(\boldsymbol{x},t)$. Hence the KG equation remains invariant under the $\mathcal{PT}$ operation, Equations (3.14)-(3.17). Notice further that Equation (3.18) is just the "quantized" equation of the kinematical relation for an antiparticle in SR

$$\left(\hat{E}_c - V_c\right)^2 = c^2\boldsymbol{p}_c^2 + m^2c^4 \tag{3.20}$$

which is the counterpart of Equation (3.4) for a particle. For example, a KG particle's scattering WF $\psi(\boldsymbol{x},t;E_1) \sim \mathrm{e}^{-\mathrm{i}E_1t}\,(E_1 > m)$ is attracted by an spherically symmetric potential $V(r) < 0$ and so has a positive phase-shift $\delta_1 > 0$ (in the, say, $S(l=0)$ state). Then physically, its antiparticle's WF $\psi_c(\boldsymbol{x},t;E_1^c) \sim \mathrm{e}^{\mathrm{i}E_1^ct}\,(E_1^c = E_1 > m)$ is repelled by the potential $V_c(r) = -V(r) > 0$ and has a negative phaseshift $\delta_1^c < 0$.

Note that, however, corresponding to $\psi(\boldsymbol{x},t;E_1)$, there is another negative energy particle's WF

$\psi(\boldsymbol{x},t;-E_1)\sim e^{iE_1t}$ satisfying Equation (3.5)

$$\begin{aligned}&\left(i\frac{\partial}{\partial t}-V\right)^2\psi(\boldsymbol{x},t;-E_1)\\&=(-E_1-V)^2\psi(\boldsymbol{x},t;-E_1)\\&=-c^2\nabla^2\psi(\boldsymbol{x},t;-E_1)+m^2\psi(\boldsymbol{x},t;-E_1)\end{aligned}\tag{3.21}$$

whose space-time behavior is precisely the same as the antiparticle's WF $\psi_c(\boldsymbol{x},t;E_1^c)\sim e^{iE_1^ct}$ with $E_1^c=E_1>m$

as shown by Equation (3.18) since

$(E_1+V)^2=(E_c-V_c)^2$. Thus, for avoiding confusion, we have

$$\begin{aligned}&\mathcal{PT}\psi(\boldsymbol{x},t;E_1)\\&=\psi(\boldsymbol{x},t;-E_1)=\mathcal{C}\psi(\boldsymbol{x},t;E_1)\\&=\psi_c(\boldsymbol{x},t;E_1^c)\neq\psi(-\boldsymbol{x},-t;E_1)\end{aligned}\tag{3.22}$$

and

$$\begin{aligned}&\mathcal{PT}\psi(\boldsymbol{x},t)\\&=\mathcal{C}\psi(\boldsymbol{x},t)=\psi(-\boldsymbol{x},-t)\\&=\psi_c(\boldsymbol{x},t)(V=0)\end{aligned}\tag{3.23}$$

achieving the proof of the discrete symmetry $\mathcal{PT}=\mathcal{C}$ for KG particle shown by Equation (3.17). In summary, the "negative-energy paradox" for KG equation is solved in a physical way with following advantages:

a) By using two sets of WFs and momentum-energy operators for particle and antiparticle respectively, both particle's WF $\psi(\boldsymbol{x},t)$ and antiparticle's WF $\psi_c(\boldsymbol{x},t)$ have positive energies $E>0$ and $E_c>0$ respectively.

b) While satisfying the same KG equation with same potential $V(r)$ formally, $\psi(x,t)$ and $\psi_c(x,t)$ are actually subject to opposite "force" for particle and antiparticle respectively.

c) The space-time behavior of $\psi_c(x,t;E_1^c)$ can be identified with that of a negative energy particle's WF $\psi(x,t;-E_1)(E_1=E_1^c)$, in a one-to-one correspondence.

Thus from mathematical point of view, all solutions of KG equation form a complete set including both positive and negative energy values of one operator $\hat{E}=i\frac{\partial}{\partial t}$ exactly.

By contrast, usually, aiming at finding an anti-particle's WF, one performs the CPT transformation on a particle's WF $\psi(x,t)$, yielding [29-32]

$$\psi(x,t)\to CPT\psi(x,t)=\psi(-x,-t) \tag{3.24}$$

whose character can also be summed up as follows:

a') By using one set of WF and relevant operators for both particle and antiparticle, at the LHS of Equation (3.24), $\psi(x,t)$, and $\psi(-x,-t)$ at RHS must have opposite energies inevitably.

b') By design in the C transformation, $\psi(x,t)$ and $\psi(-x,-t)$ in Equation (3.24) satisfy different equations with V and $V_c=-V$ respectively. But with opposite energies, they are actually subject to the same (either attractive or repulsive) "force". So one cannot distinguish particle from antiparticle through what their WFs "feel" after the CPT transformation.

c') From mathematical point of view, we should keep all negative-energy solutions for one equation. However, even facing WFs in doubled numbers, we still don't know how to choose half of them for describing particle and its antiparticle separately in physics.

But we haven't solve the difficulty of negative probability density in KG equation yet, awaiting for another enlightenment which was already there since 1958.

3.3. Feshbach and Villars (FV) Dissociation of KG WF $\psi = \phi + \chi$, a Reformulated Symmetry between ϕ and χ under the Space-Time (or Mass) Inversion

In 1958, dividing the WF into $\psi = \phi + \chi$, Feshbach and Villars [2] recast Equation (5) into two coupled Schrödinger-like equations as[3]:

$$\begin{cases} \left(i\hbar \dfrac{\partial}{\partial t} - V \right)\phi = mc^2\phi - \dfrac{\hbar^2}{2m}\nabla^2(\phi + \chi) \\ \left(i\hbar \dfrac{\partial}{\partial t} - V \right)\chi = -mc^2\chi + \dfrac{\hbar^2}{2m}\nabla^2(\phi + \chi) \end{cases} \tag{3.25}$$

where

$$\begin{cases} \phi = \dfrac{1}{2}\left[\left(1 - \dfrac{1}{mc^2}V \right)\psi + \dfrac{i\hbar}{mc^2}\dot{\psi} \right] \\ \chi = \dfrac{1}{2}\left[\left(1 + \dfrac{1}{mc^2}V \right)\psi - \dfrac{i\hbar}{mc^2}\dot{\psi} \right] \end{cases} \tag{3.26}$$

$\left(\dot{\psi} = \dfrac{\partial \psi}{\partial t} \right)$. Interestingly, the "probability density", Equation (3.11) can be recast into a difference between two positive-definite densities [18,20]:

$$\rho = \phi^*\phi - \chi^*\chi \tag{3.27}$$

while the probability current density contains interference terms between ϕ and χ:

$$\begin{aligned} \boldsymbol{j} = \frac{i\hbar}{2m}\Big[&\left(\phi\nabla\phi^* - \phi^*\nabla\phi\right) + \left(\chi\nabla\chi^* - \chi^*\nabla\chi\right) \\ &+ \left(\phi\nabla\chi^* - \chi^*\nabla\phi\right) + \left(\chi\nabla\phi^* - \phi^*\nabla\chi\right) \Big] \end{aligned} \tag{3.28}$$

The expression of ρ as shown by Equation (3.27) strongly hints that the $\mathcal{PT} = \mathcal{C}$ symmetry proved in the last subsection may be combined with the FV dissociation of KG equation such that the positive-definite property of ρ can be ensured for both particle and antiparticle.

Indeed, after inspecting Equation (3.25) carefully, we do find a hidden symmetry in the sense that it is invariant (in its form) under the following reformulated space-time inversion $(x \to -x, t \to -t)$, i.e., $\mathcal{PT} = \mathcal{C}$ transformation:

$$\begin{cases} x \to -x, t \to -t, \\ V(x,t) \to -V(x,t) = V_c(x,t), \\ \psi(x,t) \to \mathcal{PT}\psi(x,t) = \psi_c(x,t), \\ \phi(x,t) \to \mathcal{PT}\phi(x,t) = \chi_c(x,t), \\ \chi(x,t) \to \mathcal{PT}\chi(x,t) = \phi_c(x,t) \end{cases} \tag{3.29}$$

Performing transformation Equation (3.29) on Equation (3.26), we find χ_c satisfying the same equation of χ and ϕ_c satisfying that of ϕ. They read

$$\begin{cases} \chi_c = \frac{1}{2}\left[\left(1 + \frac{1}{mc^2}V\right)\psi_c - \frac{i\hbar}{mc^2}\dot{\psi}_c\right] \\ \phi_c = \frac{1}{2}\left[\left(1 - \frac{1}{mc^2}V\right)\psi_c + \frac{i\hbar}{mc^2}\dot{\psi}_c\right] \end{cases} \tag{3.30}$$

Remember, for ψ_c, we should use operator Equation (3.15). Accordingly, the probability density for ψ_c is defined as

$$\begin{aligned} \rho \to \mathcal{PT}\rho &= \rho_c \\ &= \frac{i\hbar}{2mc^2}\left(\psi_c\dot{\psi}_c^* - \psi_c^*\dot{\psi}_c\right) + \frac{1}{mc^2}V\psi_c^*\psi_c \\ &= \chi_c^*\chi_c - \phi_c^*\phi_c \end{aligned} \tag{3.31}$$

Similarly, we have $(\nabla\psi \to -\nabla\psi_c)$

$$j \to \mathcal{PT}j = j_c = \frac{i\hbar}{2m}\left(\psi_c^*\nabla\psi_c - \psi_c\nabla\psi_c^*\right) \tag{3.32}$$

For simplicity, consider a free KG particle $(V=0)$ with WF Equation (3.6). Then $|\phi|>|\chi|$

$$\begin{cases} \phi = \dfrac{1}{2}\left(1+\dfrac{E}{mc^2}\right)\psi \\ \chi = \dfrac{1}{2}\left(1-\dfrac{E}{mc^2}\right)\psi \end{cases}, \begin{cases} \rho = |\phi|^2 - |\chi|^2 > 0 \\ \boldsymbol{j} = \dfrac{1}{m}\boldsymbol{p}|\psi|^2 \end{cases} \tag{3.33}$$

But for a free $(V=0)$ KG antiparticle with WF Equation (2.21), it has $|\chi_c|>|\phi_c|$

$$\begin{cases} \phi_c = \dfrac{1}{2}\left(1-\dfrac{E_c}{mc^2}\right)\psi_c \\ \chi_c = \dfrac{1}{2}\left(1+\dfrac{E_c}{mc^2}\right)\psi_c \end{cases}, \begin{cases} \rho_c = |\chi_c|^2 - |\phi_c|^2 > 0 \\ \boldsymbol{j}_c = \dfrac{1}{m}\boldsymbol{p}_c|\psi_c|^2 \end{cases} \tag{3.34}$$

Equations (3.33) and (3.34) satisfy all physical conditions we need. If $V \neq 0$, as long as $(E-V)>0$ for particle or $(E_c-V_c)>0$ for antiparticle, the situation remains the same. However, once $(E-V)<0$ or $(E_c-V_c)<0$, some complications would occur. For further discussion, please see the Appendix.

Therefore, we see that the reformulated space-time inversion, Equation (3.29), reflects the underlying symmetry between a particle's WF ψ and its antiparticle's WF ψ_c. As both E and ρ in ψ or E_c and ρ_c in ψ_c are positive definite, all difficulties in KG equation disappear and the latter becomes a self-consistent theory.

Moreover, instead of Equation (3.29), a "mass inversion $(m \to -m)$" can realize the same symmetry, the invariance under a $\mathcal{PT}=\mathcal{C}$ transformation, via the following operation on Equation (3.25):

$$\begin{cases} m \to -m_c = -m \\ V(\boldsymbol{x},t) \to V(\boldsymbol{x},t) = -V_c(\boldsymbol{x},t), \\ \psi(\boldsymbol{x},t) \to \psi_c(\boldsymbol{x},t), \\ \phi(\boldsymbol{x},t) \to \chi_c(\boldsymbol{x},t), \\ \chi(\boldsymbol{x},t) \to \phi_c(\boldsymbol{x},t) \end{cases} \tag{3.35}$$

Notice that, when $m \to -m$, we have $\hat{\boldsymbol{p}} \to -\hat{\boldsymbol{p}}_c$ and $\hat{E} \to -\hat{E}_c$, i.e. $-\mathrm{i}\hbar\nabla \to -\mathrm{i}\hbar\nabla$, $\mathrm{i}\hbar\frac{\partial}{\partial t} \to \mathrm{i}\hbar\frac{\partial}{\partial t}$, in contrast to Equation (3.15) [1].[4]

The reason why $V \to -V$ in the space-time inversion Equation (3.29) whereas $V \to V$ in the mass inversion Equation (3.35) can be seen from the classical equation: The Lorentz force F on a particle exerted by an external potential Φ reads: $\boldsymbol{F} = -\nabla V = -\nabla(q\Phi) = m\boldsymbol{a}$. As the acceleration $\boldsymbol{a}$ of particle will change to $-\boldsymbol{a}$ for its antiparticle, there are two alternative explanations: either due to the inversion of charge $q \to -q$ (i.e., $V \to -V$ but keeping m unchanged) or due to the inversion of mass $m \to -m$ (but keeping V unchanged).

4. Reinterpretation of WF and the Relativistic Effects

The success of FV's dissociation of KG equation should be ascribed to their deep insight that a unified WF ψ is composed of two fields ϕ and χ in confrontation. Note that Equation (3.25) reduces into two equations separately for a static KG particle $(V = 0, \hbar = c = 1)$:

$$\begin{cases} \mathrm{i}\dfrac{\partial}{\partial t}\phi = m\phi, \\ \mathrm{i}\dfrac{\partial}{\partial t}\chi = -m\chi \end{cases} \tag{4.1}$$

with two separated solutions being:

$$\begin{cases} E = m > 0, \\ \phi \sim e^{-iEt}, \\ \chi = 0 \end{cases}, \quad \begin{cases} E = -m = -E_c < 0, E_c = m > 0 \\ \chi_c \sim e^{iE_c t}, \\ \phi_c = 0 \end{cases} \tag{4.2}$$

Once the particle (antiparticle) is moving with a velocity, $v \neq 0$, ϕ and χ (χ_c and ϕ_c) couple together and the WF $\psi = \phi + \chi$ ($\psi_c = \phi_c + \chi_c$) for a free particle (antiparticle) read (in one-dimensional space)

$$\psi \sim \phi \sim \chi \sim \exp[i(px - Et)], (|\phi| > |\chi|) \tag{4.3a}$$

$$\psi_c \sim \chi_c \sim \phi_c \sim \exp[-i(p_c x - E_c t)], (|\chi_c| > |\phi_c|) \tag{4.3b}$$

$(p_c = p > 0, E_c = E > 0)$ respectively. In Equation (4.3a), ϕ dominates $\chi (|\phi| > |\chi|)$. By contrast, in Equation (4.3b) it is χ_c who dominates ϕ_c (The status remains the same for $V \neq 0$ cases as discussed in the last section).

Despite ϕ and ϕ_c (χ and χ_c) having the "intrinsic tendency" to evolve as $\exp[i(px - Et)](\exp[-i(px - Et)])$, however, in a WF of particle (antiparticle), $\chi(\phi_c)$ must follow $\phi(\chi_c)$ to evolve like that shown by Equation (4.3a) (Equation

(4.3b)), as $|\phi| > |\chi| (|\chi_c| > |\phi_c|)$. So it seems suitable to name ϕ the "hidden particle field" inside a particle while χ the "hidden antiparticle field" (rather than the "negative-energy component") inside the same particle.

Let us try to reinterpret the phenomena displayed in the kinematics of special relativity (SR) via the enhancement of χ field in a particle [22-25]:

(a) Lorentz transformation Consider a particle's WF shown by Equation (4.3a) in an inertial frame S (laboratory). Then take another S' frame resting on the particle, so $p'=0$ and $E'=E_0=mc^2$. The WF in S' frame reads:

$$\psi(x',t')\sim\exp\left[\frac{\mathrm{i}}{\hbar}(p'x'-E't')\right]=\exp\left[-\frac{\mathrm{i}}{\hbar}E_0t'\right] \tag{4.4}$$

Here the space-time coordinates (x',t') are introduced and defined in the S' frame via the phase of WF as follows: Based on the assertion that "phase remains invariant under the coordinate transformation" which was named the "law of phase harmony" by de Broglie and was regarded by himself as the fundamental achievement all his life [34], comparing the phase in Equation (4.4) with that in Equation (4.3a) and using

$E=E_0\big/\sqrt{1-v^2/c^2}\,,p=Ev/c^2$, one finds

$$t'=\frac{t-vx/c^2}{\sqrt{1-v^2/c^2}} \tag{4.5}$$

Then, all formulas in the Lorentz transformation can be obtained. In some sense, what used here is a particle's wave-packet which serves as a microscopic "ruler", also a "clock" simultaneously.

(b) There is a speed limit c for a massive particle.

For a free KG particle, using Equation (3.33), we may define an "impurity ratio" R for the amplitude of hidden χ field to that of ϕ field and calculate it being

$$R_{\text{free}}^{\text{KG}}=\frac{|\chi|}{|\phi|}=\left[\frac{1-\sqrt{1-(v/c)^2}}{1+\sqrt{1-(v/c)^2}}\right] \tag{4.6}$$

When $v\to 0,|\chi|\to 0$, with the increase of v, $|\chi|/|\phi|$ increases monotonously. The particle becomes more and more "impure" until $|\chi|/|\phi|\to 1$ as a limit of particle being still a particle. As shown by Equation (4.6), the reason why its velocity has a limiting value c (the speed of light) is because ϕ and χ have opposite evolution tendencies in space-time as shown by Equations (4.1)-

(4.3) essentially, χ strives to hold ϕ back from going forward until a balance nearly reached when $|\chi| \to |\phi|$ and $v \to c$.

(c) The "length contraction" (FitzGerald-Lorentz contraction) and "time dilation"

As usual, we will show "length contraction" via a wave-packet of KG particle moving at a high-speed (v) but further ascribe it to the enhancement of χ field hidden inside the particle.

First, consider a wave-packet of KG particle at rest [25, 35]

$$\psi(x,t) = \left(4\sigma\pi^3\right)^{-1/4} \int_{-\infty}^{\infty} \exp\left(-\frac{k^2}{2\sigma}\right) \exp\left[i(kx-\omega t)\right] dk \tag{4.7}$$

Assuming $\sqrt{\sigma} \ll \frac{mc}{\hbar}$, we have approximately that

$$\psi(x,t) \cong \frac{(\sigma/\pi)^{1/4}}{(1+i\sigma\hbar t/m)^{1/2}} \exp\left[-\frac{\sigma x^2}{2(1+i\sigma\hbar t/m)} - \frac{imc^2 t}{\hbar}\right] \tag{4.8}$$

If $\sigma\hbar t/m \ll 1$, the diffusion of wave-packet at low speed $(v \ll c)$ can be ignored. Then we perform a "boost transformation"

$$\left(x \to (x-vt)/\sqrt{1-\beta^2}, t \to (t - vx/c^2)/\sqrt{1-\beta^2}, \beta = v/c\right)$$

to push the wave-packet to high velocity $(v \to c)$, yielding

$$\psi_{\text{boost}}(x,t) = \left(\frac{\sigma}{\pi}\right)^{1/4} e^{i\alpha\xi} \cdot \exp\left(-i\frac{mc^2}{\hbar}\sqrt{1-\beta^2}\, t\right) \exp\left(-\frac{\xi^2}{2\varpi^2}\right) \tag{4.9}$$

where $\xi = \frac{mc}{\hbar}(x-vt), \alpha = \beta/\sqrt{1-\beta^2}$ and

$$\varpi = \frac{mc\sqrt{1-\beta^2}}{\hbar\sqrt{\sigma}} \propto \sqrt{1-\beta^2} \tag{4.10}$$

Here ϖ is the width of wave-packet measured from its center $\xi = 0$. Equations (4.7)-(4.10) show the "length contraction".

Second, we calculate from Equations (4.9) and (3.33)

the values of $|\phi|^2, |\chi|^2$ and the probability density

$\rho = |\phi|^2 - |\chi|^2$ respectively.[5] Their peak values all increase with the increase of v (boost effect). However, the "intensity" of $|\phi|^2$ or $|\chi|^2$ increases even faster than that of ρ while keeping the constraint $|\phi| > |\chi|$ in the boosting process.

We also calculate the square of "impurity ratio" R for this moving wave-packet:

$$\left[R_{\text{free}}^{KG}\right]^2 = \frac{\int_{-\infty}^{\infty} |\chi|^2 \, dx}{\int_{-\infty}^{\infty} |\phi|^2 \, dx} = \left[\frac{1-\sqrt{1-(v/c)^2}}{1+\sqrt{1-(v/c)^2}}\right]^2 \tag{4.11}$$

which is the counterpart of Equation (4.6) for a plane WF of KG particle.

With these calculations, we might intuitively understand the length contraction as an effect of coupling (i.e. entanglement) between ϕ and χ fields due to their opposite evolution tendencies in space as discussed in previous point (b).

Let's turn to the "time dilation" shown by the variation of the mean life

$$\tau = \frac{\tau_0}{\sqrt{1-\beta^2}} \tag{4.12}$$

of a particle, say, a pion (π^- or π^+) with its velocity v.

To understand it, let's return back to Equations (4.1)- (4.3) at $x=0$ and view the WF $\psi(\psi_c)$ on its complex plane with $\mathrm{Re}\psi$ and $\mathrm{Im}\psi$ ($\mathrm{Re}\psi_c$ and $\mathrm{Im}\psi_c$) as abscissa and ordinate. We may see that the time reading of the "inner clock" for a particle (or an antiparticle) is "clockwise" (or "counter clockwise"). Thus with the increase of particle velocity, though the time reading remains clockwise (due to the dominance of ϕ field), it runs slower and slower because of the enhancement of hidden χ field.

(d) WF's group velocity u_g versus phase velocity u_p.

In RQM, a particle's velocity v should be identified with its group velocity u_g. Actually, we have

$$u_g = \frac{\mathrm{d}\omega}{\mathrm{d}k} = \frac{\mathrm{d}E}{\mathrm{d}p} = \frac{\mathrm{d}}{\mathrm{d}p}\sqrt{p^2c^2+m^2c^4}$$
$$= \frac{pc^2}{E} = v \xrightarrow[E\to\infty]{} c \tag{4.13}$$

However, the fact that there is an upper bound for particle's velocity doesn't mean that no speed can exceed that of light, c. Indeed, there is another velocity u_p, the phase velocity in the WF

$$u_p = \frac{\omega}{k} = \frac{E}{p} \tag{4.14}$$

And the relation $E^2 = p^2c^2 + m^2c^4$ implies that

$$u_g u_p = c^2, u_p = \frac{c^2}{u_g} = \frac{c^2}{v} \tag{4.15}$$

In our opinion, the role of $u_p > c$ here is crucial to maintain the quantum coherence of WF in the space-time globally, we will further discuss this problem elsewhere. In 1923, de Broglie discovered Equation (4.15) in his relativistic theory. However, in the Schrödinger equation of NRQM, the phase velocity remains undefined. See Ref [34].

5. Dirac Equation as Coupled Equations of Two-Component Spinors

Let us turn to the Dirac equation describing an electron

$$\left(\mathrm{i}\hbar\frac{\partial}{\partial t}-V\right)\psi = H\psi = \left(-\mathrm{i}\hbar c\alpha\cdot\nabla + \beta mc^2\right)\psi \tag{5.1}$$

with $\boldsymbol{\alpha}$ and β being 4×4 matrices, the WF ψ is a four-component spinor

$$\psi = \begin{pmatrix}\phi \\ \chi\end{pmatrix} \tag{5.2}$$

Usually, the two-component spinors ϕ and χ are called "positive" and "negative" energy components. In our point of view, they are the hiding "particle" and "antiparticle" fields in a particle (electron) respectively ([25], see below). Substitution of Equation (2) into Equation (1) leads to

$$\begin{cases}\left(\mathrm{i}\hbar\frac{\partial}{\partial t}-V\right)\phi = -\mathrm{i}\hbar c\boldsymbol{\sigma}\cdot\nabla\chi + mc^2\phi \\ \left(\mathrm{i}\hbar\frac{\partial}{\partial t}-V\right)\chi = -\mathrm{i}\hbar c\boldsymbol{\sigma}\cdot\nabla\phi - mc^2\chi\end{cases} \tag{5.3}$$

($\boldsymbol{\sigma}$ are Pauli matrices). Equation (3) is invariant under the combined space-time inversion with

$$\begin{cases}\boldsymbol{x}\to-\boldsymbol{x}, t\to -t, \\ \phi(\boldsymbol{x},t)\to C\phi(\boldsymbol{x},t) = \chi_c(\boldsymbol{x},t), \\ \chi(\boldsymbol{x},t)\to C\chi(\boldsymbol{x},t) = \phi_c(\boldsymbol{x},t) \\ V(\boldsymbol{x},t)\to -V(\boldsymbol{x},t) = V_c(\boldsymbol{x},t)\end{cases} \tag{5.4}$$

showing that in its form of two-component spinors, Dirac equation is in conformity with the underlying symmetry Equation (3.29). Note that under the space-time inversion, the $\boldsymbol{\sigma}$ remain unchanged (However, see Equations (9)- (11) below). Alternatively, Equation (3) also remains invariant under a mass inversion as

$$m \to -m, \phi(\boldsymbol{x},t) \to \chi_c(\boldsymbol{x},t),$$
$$\chi(\boldsymbol{x},t) \to \phi_c(\boldsymbol{x},t), V \to V \tag{5.5}$$

In either case of Equation (5.4) or (5.5), we have[6]

$$\psi(\boldsymbol{x},t) = \begin{pmatrix} \phi(\boldsymbol{x},t) \\ \chi(\boldsymbol{x},t) \end{pmatrix} \to \begin{pmatrix} \chi_c(\boldsymbol{x},t) \\ \phi_c(\boldsymbol{x},t) \end{pmatrix} = \psi_c'(\boldsymbol{x},t) \tag{5.6}$$

For concreteness, we consider a free electron moving along the z axis with momentum $p = p_z > 0$ and having a helicity $h = \boldsymbol{\sigma} \cdot \boldsymbol{p} / |\boldsymbol{p}| = 1$, its WF reads:

$$\psi(z,t) \sim \begin{pmatrix} \phi \\ \chi \end{pmatrix} \sim \begin{pmatrix} 1 \\ 0 \\ \dfrac{p}{E+m} \\ 0 \end{pmatrix} \exp\left[\mathrm{i}(pz - Et)\right] \tag{5.7}$$

with $|\phi| > |\chi|$. Under a space-time inversion

$(z \to -z, t \to -t, p \to p_c, E \to E_c)$ or mass inversion

$(m \to -m, p \to -p_c, E \to -E_c)$, it is transformed into a WF for positron (moving along z axis)

$$\psi_c'(z,t) \sim \begin{pmatrix} \chi_c \\ \phi_c \end{pmatrix} \sim \begin{pmatrix} 1 \\ 0 \\ \dfrac{p_c}{E_c+m} \\ 0 \end{pmatrix} \exp\left[-\mathrm{i}(p_c z - E_c t)\right] \tag{5.8}$$

with $|\chi_c| > |\phi_c|, (p_c > 0, E_c > 0)$. However, the positron's helicity becomes $h_c = \frac{\boldsymbol{\sigma}_c \cdot \boldsymbol{p}_c}{|\boldsymbol{p}_c|} = -1$. This is because the total angular momentum operator for an electron reads

$$\hat{\boldsymbol{J}} = \hat{\boldsymbol{L}} + \frac{\hbar}{2}\boldsymbol{\sigma} \tag{5.9}$$

Under a space-time inversion, the orbital angular momentum operator is transformed as

$$\begin{aligned} \hat{\boldsymbol{L}} &= \boldsymbol{r} \times \hat{\boldsymbol{p}} = \boldsymbol{r} \times (-\mathrm{i}\hbar\nabla) \\ &\to -\boldsymbol{r} \times (\mathrm{i}\hbar\nabla) = -\boldsymbol{r} \times \hat{\boldsymbol{p}}_c = -\hat{\boldsymbol{L}}_c \end{aligned} \tag{5.10}$$

To get $\hat{\boldsymbol{J}} \to -\hat{\boldsymbol{J}}_c$ with $\hat{\boldsymbol{J}}_c = \hat{\boldsymbol{L}}_c + \frac{\hbar}{2}\hat{\boldsymbol{\sigma}}_c$, we should have

$$\hat{\boldsymbol{\sigma}}_c = -\hat{\boldsymbol{\sigma}} \tag{5.11}$$

Hence the values of matrix element for positron's spin operator $\boldsymbol{\sigma}_c$ is just the negative to that for $\boldsymbol{\sigma}$ in the same matrix representation.

Notice that Equation (7) describes an electron with positive helicity, i.e., $\Sigma \cdot \hat{\boldsymbol{p}}\psi = p_z\psi = p\psi$ [7]. Under a space-time inversion, it is transformed into

$(-\Sigma_c) \cdot \hat{\boldsymbol{p}}_c \psi_c' = \Sigma_z \left(\mathrm{i}\hbar \frac{\partial}{\partial z}\right)\psi_c' = p_c \psi_c'$ in Equation (8), i.e.,

$\Sigma_c \cdot \hat{\boldsymbol{p}}_c \psi_c' = -p_c \psi_c'$, meaning that Equation (8) describes a positron with negative helicity.

In its form of four-component spinor, Dirac equation, Equation (5.1) with $V = 0$, is usually written in a covariant form as (Pauli metric is used: $x_4 = \mathrm{i}ct, \gamma_k = -\mathrm{i}\beta\alpha_k, \gamma_4 = \beta, \gamma_5 = \gamma_1\gamma_2\gamma_3\gamma_4 = -\begin{pmatrix} 0 & I \\ I & 0 \end{pmatrix}$, see Ref. [25]):

$$\left(\gamma_\mu \partial_\mu + m\right)\psi = 0 \tag{5.12}$$

Under a space-time (or mass) inversion, it turns into an equation for antiparticle:

$$\left(-\gamma_\mu \partial_\mu + m\right)\psi_c' = 0 \tag{5.13}$$

with an example of ψ_c' shown in Equation (8). Let us perform a representation transformation:

$$\psi_c' \to \psi_c = \left(-\gamma_5\right)\psi_c' = \begin{pmatrix} \phi_c \\ \chi_c \end{pmatrix} \tag{5.14}$$

and arrive at

$$\left(\gamma_\mu \partial_\mu + m\right)\psi_c = 0 \tag{5.15}$$

due to $\{\gamma_5, \gamma_\mu\} = 0$. Since ψ_c and ψ_c' are essentially the same in physics, (this is obviously seen from its resolved form, Equation (5.3)), it is merely a trivial thing to change the position of χ_c in the 4-component spinor (lower in Equation (5.14) and upper in Equation (5.8)).

What important is $|\chi_c| > |\phi_c|$ for characterizing an antiparticle versus $|\phi| > |\chi|$ for a particle. Therefore, if a particle with energy E runs into a potential barrier $V = V_0 > E + m$, its kinetic energy

$\left(T = E - V_0 < 0\right)$ becomes negative, and its WF's third component in Equation (5.7) suddenly turns into

$\dfrac{p'}{E - V_0 + m} = \dfrac{-p'}{V_0 - E - m}, \left(p' = \sqrt{\left(E - V_0\right)^2 - m^2}\right)$, whose absolute magnitude is larger than that of the first component. This means that it is an antiparticle's WF satisfying Equation (5.15) (with $E_c = V_0 - E(> m)$ and $|\chi_c| > |\phi_c|$) and will be crucial for the explanation of Klein paradox in Dirac equation (For detail, please see Appendix).

However, we need to discuss the "probability density" ρ and "probability current density" j for a Dirac particle versus ρ_c and j_c for its antiparticle. Different from that in KG equation, now we have

$$\rho = \psi^\dagger \psi = \phi^\dagger \phi + \chi^\dagger \chi$$
$$\rightarrow \rho_c = \psi_c^\dagger \psi_c = \chi_c^\dagger \chi_c + \phi_c^\dagger \phi_c \tag{5.16}$$

which is positive definite for either particle or antiparticle. On the other hand, we have

$$\boldsymbol{j} = c\psi^\dagger \boldsymbol{\alpha} \psi = c\left(\phi^\dagger \boldsymbol{\sigma} \chi + \chi^\dagger \boldsymbol{\sigma} \phi\right)$$
$$\rightarrow \boldsymbol{j}_c = c\psi_c^\dagger \boldsymbol{\alpha} \psi_c = c\left(\chi_c^\dagger \boldsymbol{\sigma} \phi_c + \phi_c^\dagger \boldsymbol{\sigma} \chi_c\right) \tag{5.17}$$

(we prefer to keep $\boldsymbol{\sigma}$ rather than $\boldsymbol{\sigma}_c$ for antiparticle). For Equations (5.7), (5.8) and (5.14), we find $(c = \hbar = 1)$

$$j_z \sim \frac{2p}{E+m} > 0 \rightarrow j_z^c \sim \frac{2p_c}{E_c + m} > 0\left(V = 0\right) \tag{5.18}$$

which means that the probability current is always along the momentum's direction for either a particle or antiparticle.

Above discussions at RQM level may be summarized as follows: The first symptom for the appearance of an antiparticle is: If we perform an energy operator $\left(E = i\hbar\partial/\partial t\right)$ on a WF and find a negative energy $(E<0)$ or a negative kinetic energy $\left(E - V < 0\right)$, we'd better doubt the WF being a description of antiparticle and use the operators for antiparticle, Equation (2.18). Then for further confirmation, two more criterions for ρ and $\boldsymbol{j}$ are needed (see Appendix).

6. The Strong Reflection Invariance in CPT Theorem and QFT

In QFT, the starting point is the field operator which is constructed for free complex boson field as [36]

$$\begin{cases} \hat{\psi}(\boldsymbol{x},t)=\sum_{p}\frac{1}{\sqrt{2V\omega_{p}}}\left\{\hat{a}_{p}\exp\left[\mathrm{i}(\boldsymbol{p}\cdot\boldsymbol{x}-Et)\right]+\hat{b}_{p}^{\dagger}\exp\left[-\mathrm{i}(\boldsymbol{p}\cdot\boldsymbol{x}-Et)\right]\right\} \\ \hat{\psi}^{\dagger}(\boldsymbol{x},t)=\sum_{p}\frac{1}{\sqrt{2V\omega_{p}}}\left\{\hat{a}_{p}^{\dagger}\exp\left[-\mathrm{i}(\boldsymbol{p}\cdot\boldsymbol{x}-Et)\right]+\hat{b}_{p}\exp\left[\mathrm{i}(\boldsymbol{p}\cdot\boldsymbol{x}-Et)\right]\right\} \end{cases} \tag{6.1}$$

Similarly, the field operator for free Dirac field reads:

$$\begin{cases} \hat{\psi}(\boldsymbol{x},t)=\frac{1}{\sqrt{V}}\sum_{p}\sum_{h=\pm1}\sqrt{\frac{m}{E}}\left[\hat{a}_{p}^{(h)}u^{(h)}(\boldsymbol{p})\mathrm{e}^{\mathrm{i}(\boldsymbol{p}\cdot\boldsymbol{x}-Et)}+\hat{b}_{p}^{(h)\dagger}v^{(h)}(\boldsymbol{p})\mathrm{e}^{-\mathrm{i}(\boldsymbol{p}\cdot\boldsymbol{x}-Et)}\right] \\ \hat{\psi}^{\dagger}(\boldsymbol{x},t)=\frac{1}{\sqrt{V}}\sum_{p}\sum_{h=\pm1}\sqrt{\frac{m}{E}}\left[\hat{a}_{p}^{(h)\dagger}u^{(h)\dagger}(\boldsymbol{p})\mathrm{e}^{\mathrm{i}(\boldsymbol{p}\cdot\boldsymbol{x}-Et)}+\hat{b}_{p}^{(h)}v^{(h)\dagger}(\boldsymbol{p})\mathrm{e}^{\mathrm{i}(\boldsymbol{p}\cdot\boldsymbol{x}-Et)}\right] \end{cases} \tag{6.2}$$

In Equation (6.1), the annihilation operator $\hat{a}_p$ for particle and the creation operator $\hat{b}_p^{\dagger}$ for antiparticle in Fock space are introduced. In Equation (6.2), instead of index s ($=\pm 1/2$, the spin's projection along the fixed z axis in space), the helicity h is used. See Ref. [37].

Let us return back to the CPT theorem proved by Lüders and Pauli in 1954-1957 [10-12]. The proof of CPT theorem contains a crucial step being the construction of so-called "strong reflection", consisting in a reflection of space and time about some arbitrarily chosen origin, i.e. $\boldsymbol{r} \to -\boldsymbol{r}, t \to -t$.

Pauli proposed and explained the strong reflection in Ref. [12] as follows: When the space-time coordinates change their sign, every particle transforms into its antiparticle simultaneously. The physical sense of the strong reflection is the substitution of every emission (absorption) operator of a particle by the corresponding absorption (emission) operator of its antiparticle. And there is no need to reverse the sign of the electric charge when the sign of space-time coordinates is reversed.

What Pauli claimed, in our understanding, means that under the strong reflection for boson field, one has

$$\begin{cases} \boldsymbol{x} \to -\boldsymbol{x}, t \to -t, \\ \hat{a}_p \leftrightarrows \hat{b}_p^{\dagger}, \hat{a}_p^{\dagger} \leftrightarrows \hat{b}_p \end{cases} \tag{6.3}$$

The mutual transformation, Equation (6.3), in Fock space ensures the field operators, Equation (6.1), invariant under the strong reflection in the sense of (see also [25,26]):

$$\begin{aligned}
&\hat{\psi}(\boldsymbol{x},t) \to \left(\widehat{\mathcal{PT}}\right)\hat{\psi}(\boldsymbol{x},t)\left(\widehat{\mathcal{PT}}\right)^{-1} \\
&= \hat{\psi}(-\boldsymbol{x},-t) = \hat{\psi}(\boldsymbol{x},t) \\
&\hat{\psi}^{\dagger}(\boldsymbol{x},t) \to \left(\widehat{\mathcal{PT}}\right)\hat{\psi}^{\dagger}(\boldsymbol{x},t)\left(\widehat{\mathcal{PT}}\right)^{-1} \\
&= \hat{\psi}^{\dagger}(-\boldsymbol{x},-t) = \hat{\psi}^{\dagger}(\boldsymbol{x},t)
\end{aligned} \tag{6.4}$$

Here let us introduce the notation $\widehat{\mathcal{PT}}$ to represent the strong reflection so that the presentation could be easier and clearer as shown above. Similarly, for Dirac field, under the strong reflection one has

$$\begin{cases} \boldsymbol{x} \to -\boldsymbol{x}, t \to -t, \\ \hat{a}_{\boldsymbol{p}}^{(h)} \leftrightarrows \hat{b}_{\boldsymbol{p}}^{(-h)\dagger}, \hat{a}_{\boldsymbol{p}}^{(h)\dagger} \leftrightarrows \hat{b}_{\boldsymbol{p}}^{(-h)} \end{cases} \tag{6.5}$$

Here it is important to notice that the helicity, h, will be reversed before and after the strong reflection for a particle and its antiparticle respectively as discussed in Section V. Because Equation (6.2) is written in 4 component spinor covariant form, the invariance of Dirac field operator under the strong reflection should be expressed rigorously as

$$\begin{aligned}
&\hat{\psi}(\boldsymbol{x},t) \to \left(\widehat{\mathcal{PT}}\right)\hat{\psi}(\boldsymbol{x},t)\left(\widehat{\mathcal{PT}}\right)^{-1} \\
&= -\gamma_5\hat{\psi}(-\boldsymbol{x},-t) = \hat{\psi}(\boldsymbol{x},t) \\
&\hat{\psi}^{\dagger}(\boldsymbol{x},t) \to \left(\widehat{\mathcal{PT}}\right)\hat{\psi}^{\dagger}(\boldsymbol{x},t)\left(\widehat{\mathcal{PT}}\right)^{-1} \\
&= \hat{\psi}^{\dagger}(-\boldsymbol{x},-t)(-\gamma_5) = \hat{\psi}^{\dagger}(\boldsymbol{x},t)
\end{aligned} \tag{6.6}$$

$$\begin{aligned}
&\hat{\psi}(-\boldsymbol{x},-t) = -\gamma_5\hat{\psi}(\boldsymbol{x},t), \\
&\hat{\psi}^{\dagger}(-\boldsymbol{x},-t) = \hat{\psi}^{\dagger}(\boldsymbol{x},t)(-\gamma_5)
\end{aligned} \tag{6.7}$$

which are useful in proving the “spin-statistics connection” by strong reflection invariance.

QFT is a successful theory just because it is established on sound basis with the field operator being one of its cornerstones. Historically, through various trials and checks, Equations (6.1)-(6.2) were eventually found (see Section 3.5 of Ref. [36]). Why they are correct and why one would fail otherwise? In our understanding, it is just because they are invariant under the strong reflection as shown by Equations (6.4) and (6.6).

However, as emphasized by Pauli [12] and further stressed by Lüders [11], at least two more rules should be added in doing calculations:

(a) The order of an operator product in Fock space has to be reversed under the strong reflection, e.g.,

$\left(\widehat{PT}\right)\hat{A}\hat{B}\left(\widehat{PT}\right)^{-1}=\left(\widehat{PT}\right)\hat{B}\left(\widehat{PT}\right)^{-1}\left(\widehat{PT}\right)\hat{A}\left(\widehat{PT}\right)^{-1}$. So is the order of a process occurred in a many-particle system.

(b) Another rule is: One should always take the normal ordering when dealing with quadratic forms like $\hat{\bar{\psi}}(x)\hat{\psi}(x)$ etc.

Then Pauli and Lüders were able to prove that the Hamiltonian density $\mathcal{H}(\boldsymbol{x},t)$ for a broad kind of model in relativistic QFT is invariant under an operation of "strong reflection", i.e.,

$$\begin{aligned}\hat{\mathcal{H}}(\boldsymbol{x},t)&\to\widehat{PT}\hat{\mathcal{H}}(\boldsymbol{x},t)\left(\widehat{PT}\right)^{-1}\\&=\hat{\mathcal{H}}(-\boldsymbol{x},-t)=\hat{\mathcal{H}}(\boldsymbol{x},t)\end{aligned} \tag{6.8}$$

The Hamiltonian density is also invariant under a Hermitian conjugation (H.C.) as:

$$\hat{\mathcal{H}}(\boldsymbol{x},t)\to\hat{\mathcal{H}}^{\dagger}(\boldsymbol{x},t)=\hat{\mathcal{H}}(\boldsymbol{x},t) \tag{6.9}$$

Furthermore, they proved the CPT theorem via the identification of the product of T, C, and P in QFT with the combined operation of the strong reflection and a Hermitian conjugation.

The validity of CPT invariance, i.e. Equations (6.8) and (6.9) has been verified experimentally since the discovery of parity violation ([3-8] etc.) and the establishment (and development) of standard model ([38] etc.) in particle physics till this day. See the excellent book, Ref. [19] and the Review of Particle Physics, Ref. [9].

After restudying the historical contribution of PauliLüders strong reflection invariance, we feel good in understanding that what we claim in RQM (Sections III-V) is essentially the same as or very close to their idea.

In fact, this paper is the direct continuation of our first one in 1974 [22], which was inspired jointly by the discoveries of violations in P, C, CP, T symmetries individually (but CPT invariance holds), also by Lee-Wu's proposal in 1965 that the relationship between a particle $|a\rangle$ and its antiparticle $|\bar{a}\rangle$ should be [13]:

$$|\bar{a}\rangle = CPT|a\rangle \tag{6.10}$$

and especially by Pauli's invention of the strong reflection in 1955 [12].

Below, we would like to show that WFs for a particle and its antiparticle given in Equations (5.7) and (5.8) are precisely that derived from QFT as expected.

Using Equation (6.2) for Dirac field, we find the WF of an electron being

$$\begin{aligned}\psi_{e^-}(\boldsymbol{x},t) &= \langle 0|\hat{\psi}(\boldsymbol{x},t)|e^-,\boldsymbol{p}_1,h_1\rangle \\ &= \langle 0|\hat{\psi}(\boldsymbol{x},t)|\hat{a}^{(h_1)\dagger}_{\boldsymbol{p}_1}|0\rangle \\ &= \frac{1}{\sqrt{V}}\sqrt{\frac{m}{E_1}}u^{(h_1)}(\boldsymbol{p}_1)\mathrm{e}^{-\mathrm{i}(\boldsymbol{p}_1\cdot\boldsymbol{x}-E_1t)}\end{aligned} \tag{6.11}$$

but the hermitian conjugate of a positron's WF is given by

$$\begin{aligned}\psi^\dagger_{e^+}(\boldsymbol{x},t) &= \langle 0|\hat{\psi}^\dagger(\boldsymbol{x},t)|e^+,\boldsymbol{p}_c,h_c\rangle \\ &= \langle 0|\hat{\psi}^\dagger(\boldsymbol{x},t)|\hat{b}^{(h_c)\dagger}_{\boldsymbol{p}_c}|0\rangle \\ &= \frac{1}{\sqrt{V}}\sqrt{\frac{m}{E_c}}v^{(h_c)\dagger}(\boldsymbol{p}_c)\mathrm{e}^{\mathrm{i}(\boldsymbol{p}_c\cdot\boldsymbol{x}-E_ct)}\end{aligned} \tag{6.12}$$

which leads to positron's WF being

$$\psi_{e^+}(\boldsymbol{x},t) = \frac{1}{\sqrt{V}}\sqrt{\frac{m}{E_c}}v^{(h_c)}(\boldsymbol{p}_c)\mathrm{e}^{-\mathrm{i}(\boldsymbol{p}_c\cdot\boldsymbol{x}-E_ct)} \tag{6.13}$$

Similarly, Equations (2.20) and (2.21) can be derived from Equation (6.1) as expected.

7. An Oversight in QFT (Helicity States or Spin States?)—Why a Parity-Violation Phenomenon Was Overlooked Since 1956-1957?

Through analysis in RQM till QFT, we stress the necessity of using helicity (h) to describe a fermion or antifermion. Here is an interesting example. Since 2002, Shi and Ni [39-43] predicted a parity-violation phenomenon as follows:

An unstable (decaying) fermion (e.g., neutron or muon) has different mean lifetimes for being right-handed (RH) or left-handed (LH) polarized during its flight with the same speed $v(\beta = v/c)$

$$\tau_R = \frac{\tau}{1-\beta}, \tau_L = \frac{\tau}{1+\beta} \tag{7.1}$$

where $\tau = \tau_0 / \sqrt{1-\beta^2}$, τ_0 the mean lifetime when it is at rest. Similarly, for its antifermion, their lifetimes will be

$$\overline{\tau}_R = \frac{\tau}{1+\beta}, \overline{\tau}_L = \frac{\tau}{1-\beta} \tag{7.2}$$

Hence, the lifetime asymmetry can be defined as

$$A = \frac{\tau_R - \tau_L}{\tau_R + \tau_L} = \beta \tag{7.3}$$

This is not a small effect. For instance, in Fermilab, physicists consider to build a muon collider [44]. The collision of μ^- and μ^+ beams must happen before the muons decay. It was estimated that if a muon rings along at 1.5 TeV, the time dilation of SR stretches its lifetime to 30 milliseconds—up from 2 microseconds when it's still. That's time enough for 500 circuits in the final ring. However, as discussed in Ref. [43], if the prediction of life asymmetry Equation (7.1) is correct, the lifetime of RH μ^- will be stretched to 146 days while that of LH μ^- only 15 milliseconds. The lifetime asymmetry of μ^+ will be just the opposite as shown by Equation (7.2). Therefore, it seems necessary to take Equations (7.1)- (7.2) into account in the design of a muon collider.

The problem is: How can such a parity-violation phenomenon be overlooked since 1956-1957? One theoretical reason is: in the past, for describing a fermion in flight $(v \neq 0)$, instead of helicity states, the "spin-states" assigned by s (spin's projection along the fixed z axis in space) were often incorrectly used (see [40-42]). So previous calculations on the lifetime always led to a prediction that $\tau = \tau_0 / \sqrt{1-\beta^2}$ without parity-violation in contrast to Equations (7.1)-(7.3).[8]

The interesting thing is: While Equations (7.1) and (7.2) display the violation of P or C symmetry to its maximum, their "cross-symmetry", $\tau_R = \overline{\tau}_L$ and $\tau_L = \overline{\tau}_R$, reflects the symmetry of $\mathcal{PT} = \mathcal{C}$ shown by Equation (6.5) exactly.

8. DIRAC PARTICLES CONSERVE THE PARITY WHEREAS NEUTRINOS ARE LIKELY THE TACHYONS

8.1. Why Dirac Equation Respects the Parity Symmetry?

In the standard representation of Dirac equation for free particle

$$(\hbar = c = 1)$$

$$i\frac{\partial}{\partial t}\psi^{(D)} = -i\boldsymbol{\alpha}\cdot\nabla\psi^{(D)} + \beta m\psi^{(D)} \tag{8.1}$$

Let us choose

$$\boldsymbol{\alpha} = -\begin{pmatrix} 0 & \boldsymbol{\sigma} \\ \boldsymbol{\sigma} & 0 \end{pmatrix}, \beta = \begin{pmatrix} I & 0 \\ 0 & -I \end{pmatrix}, \psi^{(D)} = \begin{pmatrix} \phi^{(D)} \\ \chi^{(D)} \end{pmatrix}$$, then

$$\begin{cases} \mathrm{i}\frac{\partial}{\partial t}\phi^{(D)} = \mathrm{i}\boldsymbol{\sigma}\cdot\nabla\chi^{(D)} + m\phi^{(D)} \\ \mathrm{i}\frac{\partial}{\partial t}\chi^{(D)} = \mathrm{i}\boldsymbol{\sigma}\cdot\nabla\phi^{(D)} - m\chi^{(D)} \end{cases} \tag{8.2}$$

As discussed in section V, Equations (8.1) and (8.2) are invariant under the space-time inversion:

$$\begin{cases} \boldsymbol{x}\to-\boldsymbol{x}, t\to-t \\ \phi^{(D)}(\boldsymbol{x},t)\to\phi^{(D)}(-\boldsymbol{x},-t)=\chi_c^{(D)}(\boldsymbol{x},t) \\ \chi^{(D)}(\boldsymbol{x},t)\to\chi^{(D)}(-\boldsymbol{x},-t)=\phi_c^{(D)}(\boldsymbol{x},t) \end{cases} \tag{8.3}$$

with subscript "c" meaning the antiparticle.

After transforming $\psi^{(D)}$ into the "Weyl representation" (chiral representation) as

$$\psi^{(D)} \to \frac{1}{\sqrt{2}}\begin{pmatrix} I & I \\ I & -I \end{pmatrix}\begin{pmatrix} \phi^{(D)} \\ \chi^{(D)} \end{pmatrix} = \begin{pmatrix} \xi^{(D)} \\ \eta^{(D)} \end{pmatrix} \tag{8.4}$$

we have

$$\begin{cases} \mathrm{i}\frac{\partial}{\partial t}\xi^{(D)} = \mathrm{i}\boldsymbol{\sigma}\cdot\nabla\xi^{(D)} + m\eta^{(D)} \\ \mathrm{i}\frac{\partial}{\partial t}\eta^{(D)} = -\mathrm{i}\boldsymbol{\sigma}\cdot\nabla\eta^{(D)} + m\xi^{(D)} \end{cases} \tag{8.5}$$

If $m=0$, Equation (8.5) reduces into two Weyl equations describing two kinds of permanently LH and RH polarized massless fermions respectively.

So we may name $\xi^{(D)}$ and $\eta^{(D)}$ (which are usually called as chirality states or chiral fields in 4-component covariant form) as the "hidden LH and RH spinning fields" inside a Dirac particle, which can be either LH or RH polarized (with helicity $h=-1$ or 1) explicitly. See below.

A new symmetry is hidden in Equation (8.5), which remains invariant under the pure space inversion $\left(\boldsymbol{x} \to -\boldsymbol{x}, t \to t\right)$ transformation, i.e., the parity operation as

$$\begin{cases} \xi^{(D)}\left(\boldsymbol{x},t\right) \to \xi^{(D)}\left(-\boldsymbol{x},t\right) = \eta^{(D)'}\left(\boldsymbol{x},t\right) \\ \eta^{(D)}\left(\boldsymbol{x},t\right) \to \eta^{(D)}\left(-\boldsymbol{x},t\right) = \xi^{(D)'}\left(\boldsymbol{x},t\right) \end{cases} \tag{8.6}$$

Here we add " ' " in the superscript of RHS to stress that the WF after the space inversion may be different from that at the LHS (before the space inversion). We knew that the WF in Dirac representation after a space inversion reads

$$\hat{P}\psi^{(D)}\left(\boldsymbol{x},t\right) = \gamma_4 \psi^{(D)}\left(-\boldsymbol{x},t\right) \tag{8.7}$$

Using Equation (8.6), the RHS of Equation (8.7) turns out to be

$$\begin{aligned} &\frac{1}{\sqrt{2}}\gamma_4 \begin{pmatrix} \xi^{(D)}\left(-\boldsymbol{x},t\right) + \eta^{(D)}\left(-\boldsymbol{x},t\right) \\ \xi^{(D)}\left(-\boldsymbol{x},t\right) - \eta^{(D)}\left(-\boldsymbol{x},t\right) \end{pmatrix} \\ &= \frac{1}{\sqrt{2}} \begin{pmatrix} \xi^{(D)'}\left(-\boldsymbol{x},t\right) + \eta^{(D)'}\left(-\boldsymbol{x},t\right) \\ \xi^{(D)'}\left(-\boldsymbol{x},t\right) - \eta^{(D)'}\left(-\boldsymbol{x},t\right) \end{pmatrix} = \psi^{(D)'}\left(\boldsymbol{x},t\right) \end{aligned} \tag{8.8}$$

Hence, we understand the reason why a Dirac particle respects the parity symmetry as shown by Equation (8.7) is because it enjoys the symmetry Equation (8.6) hiding in the 2-component spinor form (in Weyl representation).

For concreteness, let's write down the solution of Equation (8.1)

$$\begin{aligned} &\psi^{(D)}\left(\boldsymbol{x},t\right) = \begin{pmatrix} \phi^{(D)} \\ \chi^{(D)} \end{pmatrix} \sim \begin{pmatrix} \phi_0 \\ \dfrac{-\boldsymbol{\sigma}\cdot\boldsymbol{p}}{E+m}\phi_0 \end{pmatrix}, \\ &\left(E = \sqrt{p^2 + m^2} > 0\right) \end{aligned} \tag{8.9}$$

Furthermore, we choose a simplest "spin state" with

$\hat{\boldsymbol{p}}\psi^{(D)} = p_z\psi^{(D)}$ and $\hat{\sigma}_z\psi^{(D)} = \psi^{(D)}$:

$$\psi^{(D)}_{s_z=1/2}(z,t) = \begin{pmatrix} \phi^{(D)} \\ \chi^{(D)} \end{pmatrix} \sim \begin{pmatrix} 1 \\ 0 \\ \frac{-p_z}{E+m} \\ 0 \end{pmatrix} e^{i(p_z z - Et)} \; (E>0) \tag{8.10}$$

While Equation (8.10) is an eigenfunction of $\hat{\sigma}_z$ with eigenvalue $s_z = 1/2$, its helicity h remains unfixed, depending on the value of p_z being positive or negative. Only after $p_z = p > 0$ is fixed, can we have a "helicity state" describing a RH particle with $h = 1$:

$$\psi^{(D)}_{\mathrm{RH}}(z,t) = \begin{pmatrix} \phi^{(D)} \\ \chi^{(D)} \end{pmatrix} \sim \begin{pmatrix} 1 \\ 0 \\ \frac{-p}{E+m} \\ 0 \end{pmatrix} e^{i(pz - Et)}$$

$$(p>0, E>0) \tag{8.11}$$

Looking at Equation (8.11) in the Weyl representation, we see that

$$\xi^{(D)} = \frac{1}{\sqrt{2}}\left(\phi^{(D)} + \chi^{(D)}\right) \sim \frac{1}{\sqrt{2}} \begin{pmatrix} 1 - \frac{p}{E+m} \\ 0 \end{pmatrix},$$

$$\eta^{(D)} = \frac{1}{\sqrt{2}}\left(\phi^{(D)} - \chi^{(D)}\right) \sim \frac{1}{\sqrt{2}} \begin{pmatrix} 1 + \frac{p}{E+m} \\ 0 \end{pmatrix} \tag{8.12}$$

$\left|\xi^{(D)}\right| < \left|\eta^{(D)}\right|$. So Equation (8.11) describes a RH particle just because the $\eta^{(D)}$ field dominates the $\xi^{(D)}$ field. Now we perform a space inversion on Equation (8.11), according to the rule Equation (8.7), yielding

$$\hat{P}\psi_{\mathrm{RH}}^{(D)}(z,t) \sim \begin{pmatrix} 1 \\ 0 \\ \dfrac{p}{E+m} \\ 0 \end{pmatrix} e^{i(-pz-Et)} = \begin{pmatrix} \phi^{(D)'} \\ \chi^{(D)'} \end{pmatrix} = \psi^{(D)'}(z,t)$$

$$\xi^{(D)'} = \frac{1}{\sqrt{2}}\left(\phi^{(D)'} + \chi^{(D)'}\right) \sim \frac{1}{\sqrt{2}}\begin{pmatrix} 1+\dfrac{p}{E+m} \\ 0 \end{pmatrix},$$

$$\eta^{(D)'} = \frac{1}{\sqrt{2}}\left(\phi^{(D)'} - \chi^{(D)'}\right) \sim \frac{1}{\sqrt{2}}\begin{pmatrix} 1-\dfrac{p}{E+m} \\ 0 \end{pmatrix}$$

$$\xi^{(D)'} = \frac{E+p}{m}\eta^{(D)'}, \left|\xi^{(D)'}\right| > \left|\eta^{(D)'}\right| \tag{8.13}$$

Hence we see that the reason why $\psi^{(D)'}(z,t)$ becomes a LH WF, i.e.,

$$\hat{P}\psi_{\mathrm{RH}}^{(D)}(z,t) = \psi_{\mathrm{LH}}^{(D)}(z,t) \tag{8.14}$$

is just because of the dominance of $\xi^{(D)'}$ field over $\eta^{(D)'}$ field after the P-operation. Before and after the operation, $p \to -p$, the dominant (subordinate) field is transformed into dominant (subordinate) field:

$\eta^{(D)} \to \xi^{(D)'}, \left(\xi^{(D)} \to \eta^{(D)'}\right)$, as shown by Equation (8.6).

In summary, Dirac equation is invariant under a space inversion whereas its concrete solution of WF may be not. The latter may change from that for a RH particle to a LH one or vice versa, but with the same mass m, showing the law of parity conservation exactly.

8.2. Tachyon Equation as a Counterpart of the Dirac Equation

Now a question arises: Can we find an equation which violates the symmetry of pure space inversion?

The answer is "yes". Let's introduce a new equation in Weyl representation from Equation (8.5) by erasing the superscript (D), replacing the mass term by $m \to m_s$ and changing its sign from "+" to "−" in the first equation of Equation (8.5) only [46]

$$\begin{cases} \mathrm{i}\dfrac{\partial}{\partial t}\xi = \mathrm{i}\boldsymbol{\sigma}\cdot\nabla\xi - m_s\eta \\ \mathrm{i}\dfrac{\partial}{\partial t}\eta = -\mathrm{i}\boldsymbol{\sigma}\cdot\nabla\eta + m_s\xi \end{cases} \tag{8.15}$$

where m_s (real and positive) refers to the mass of a hypothetical particle. We will see immediately that it is a "superluminal particle" or "tachyon".

Indeed, substituting a plane-wave solution

$$\xi \sim \eta \sim \exp\left[\mathrm{i}\left(p_z z - Et\right)\right]\begin{pmatrix} 0 \\ 1 \end{pmatrix} \tag{8.16}$$

with the particle's helicity $h = -1$ into Equation (8.15), we find that $\left(p_z = p > 0, E > 0\right)$

$$E^2 = p^2 - m_s^2 \tag{8.17}$$

$$\xi = \frac{1}{m_s}\left(p + E\right)\eta, |\xi| > |\eta| \tag{8.18}$$

Since $E = \hbar\omega$ and $\boldsymbol{p} = \hbar\boldsymbol{k}$, from Equation (8.17), the dispersion-relation of wave reads

$$\omega^2 = k^2 - m_s^2 \tag{8.19}$$

As in Section IV, we define the wave's phase velocity u_p as

$$u_p = \frac{\omega}{k} \tag{8.20}$$

while its group velocity u_g

$$u_g = \frac{\mathrm{d}\omega}{\mathrm{d}k} = v \tag{8.21}$$

being identical with the particle's velocity v. Equation (8.19) yields a relation between them coinciding with Equation (4.15) exactly:

$$u_p u_g = c^2 \tag{8.22}$$

However, the relations among E, p and v are dramatically different

$$E = \frac{m_s c^2}{\sqrt{\frac{v^2}{c^2} - 1}}, p = \frac{m_s v}{\sqrt{\frac{v^2}{c^2} - 1}} \tag{8.23}$$

which dictate $v > c$ such that $E, \boldsymbol{p}$ are real and $E > 0$.

Like Equation (8.4), we define:

$$\phi = \frac{1}{\sqrt{2}}(\xi + \eta), \chi = \frac{1}{\sqrt{2}}(\xi - \eta) \tag{8.24}$$

and find from Equation (8.15) that (in Dirac representation)

$$\mathrm{i}\frac{\partial}{\partial t}\psi = -\mathrm{i}\boldsymbol{\alpha} \cdot \nabla \psi + \beta_s m_s \psi \tag{8.25}$$

$$\begin{cases} \mathrm{i}\dfrac{\partial}{\partial t}\phi = \mathrm{i}\boldsymbol{\sigma}\cdot\nabla\chi + m_s\chi \\ \mathrm{i}\dfrac{\partial}{\partial t}\chi = \mathrm{i}\boldsymbol{\sigma}\cdot\nabla\phi - m_s\phi \end{cases} \tag{8.26}$$

$\left(\psi = \begin{pmatrix}\phi \\ \chi\end{pmatrix}, \beta_s = \begin{pmatrix} 0 & I \\ -I & 0\end{pmatrix}\right)$. Despite the difference between Equation (8.26) and Dirac equation, Equation (8.2), both of them respect the combined space-time inversion $(\mathcal{PT})$ symmetry like Equation (8.3)

$$\begin{cases} \phi(\boldsymbol{x},t) \to \phi(-\boldsymbol{x},-t) = \chi_c(\boldsymbol{x},t) \\ \chi(\boldsymbol{x},t) \to \chi(-\boldsymbol{x},-t) = \phi_c(\boldsymbol{x},t) \end{cases} \tag{8.27}$$

with

$$\begin{cases} \mathrm{i}\dfrac{\partial}{\partial t}\chi_c = \mathrm{i}\boldsymbol{\sigma}\cdot\nabla\phi_c - m_s\phi_c \\ \mathrm{i}\dfrac{\partial}{\partial t}\phi_c = \mathrm{i}\boldsymbol{\sigma}\cdot\nabla\chi_c + m_s\chi_c \end{cases} \tag{8.28}$$

Similarly, we define the WF in Weyl representation after $\mathcal{PT}$ inversion as:

$$\begin{cases} \xi(\boldsymbol{x},t) \to \mathcal{PT}\xi(\boldsymbol{x},t) = \xi(-\boldsymbol{x},-t) = \eta_c(\boldsymbol{x},t) \\ \eta(\boldsymbol{x},t) \to \mathcal{PT}\eta(\boldsymbol{x},t) = \eta(-\boldsymbol{x},-t) = \xi_c(\boldsymbol{x},t) \end{cases} \tag{8.29}$$

Based on Equations (8.27)-(8.29), we find

$$\begin{cases} \eta_c(\boldsymbol{x},t) = \dfrac{1}{\sqrt{2}}\left[\chi_c(\boldsymbol{x},t) + \phi_c(\boldsymbol{x},t)\right] \\ \xi_c(\boldsymbol{x},t) = \dfrac{1}{\sqrt{2}}\left[\chi_c(\boldsymbol{x},t) - \phi_c(\boldsymbol{x},t)\right] \end{cases} \tag{8.30}$$

$$\begin{cases} \mathrm{i}\dfrac{\partial}{\partial t}\eta_c = \mathrm{i}\boldsymbol{\sigma}\cdot\nabla\eta_c + m_s\xi_c \\ \mathrm{i}\dfrac{\partial}{\partial t}\xi_c = -\mathrm{i}\boldsymbol{\sigma}\cdot\nabla\xi_c - m_s\eta_c \end{cases} \tag{8.31}$$

which can also be obtained via the $\mathcal{PT}$ operation on Equation (8.15). Equations (8.15) and (8.31) are better to be compared in the following form:

$$\begin{cases} \hat{E}\xi = -\boldsymbol{\sigma}\cdot\hat{\boldsymbol{p}}\xi - m_s\eta \\ \hat{E}\eta = \boldsymbol{\sigma}\cdot\hat{\boldsymbol{p}}\eta + m_s\xi \end{cases} \tag{8.32}$$

$$\begin{cases} \hat{E}_c\eta_c = \boldsymbol{\sigma}_c\cdot\hat{\boldsymbol{p}}_c\eta_c - m_s\xi_c \\ \hat{E}_c\xi_c = -\boldsymbol{\sigma}_c\cdot\hat{\boldsymbol{p}}_c\xi_c + m_s\eta_c \end{cases} \tag{8.33}$$

$\left(\hat{E}_c = -\mathrm{i}\dfrac{\partial}{\partial t}, \hat{\boldsymbol{p}}_c = \mathrm{i}\nabla, \boldsymbol{\sigma}_c = -\boldsymbol{\sigma}\right)$. Interestingly, Equation (8.33) can also be reached from Equation (8.32) via a "mass inversion" like that in Sections III and V:

$$\begin{cases} m_s \to -m_s, \\ \hat{E} \to -\hat{E}_c = \mathrm{i}\dfrac{\partial}{\partial t}(\text{no change in } t) \\ \hat{\boldsymbol{p}} \to -\hat{\boldsymbol{p}}_c = -\mathrm{i}\nabla(\text{no change in } x) \\ \boldsymbol{\sigma} \to -\boldsymbol{\sigma}_c = \boldsymbol{\sigma}(\text{no change in } \boldsymbol{\sigma}) \\ \xi(\boldsymbol{x},t) \to \eta_c(\boldsymbol{x},t), \eta(\boldsymbol{x},t) \to \xi_c(\boldsymbol{x},t) \end{cases} \tag{8.34}$$

Furthermore, the probability density and probability current density before and after the $\mathcal{PT}$ inversion can be derived as:

$$\rho = \phi^\dagger\chi + \chi^\dagger\phi = \xi^\dagger\xi - \eta^\dagger\eta \xrightarrow[\mathcal{PT}]{} \\ \rho_c = \chi_c^\dagger\phi_c + \phi_c^\dagger\chi_c = \eta_c^\dagger\eta_c - \xi_c^\dagger\xi_c \tag{8.35}$$

and

$$\boldsymbol{j} = -\left(\phi^{\dagger}\sigma\phi + \chi^{\dagger}\sigma\chi\right) = -\left(\xi^{\dagger}\sigma\xi + \eta^{\dagger}\sigma\eta\right) \xrightarrow{\mathcal{PT}}$$
$$\boldsymbol{j}_c = -\left(\chi_c^{\dagger}\sigma\chi_c + \phi_c^{\dagger}\sigma\phi_c\right) = -\left(\eta_c^{\dagger}\sigma\eta_c + \xi_c^{\dagger}\sigma\xi_c\right) \tag{8.36}$$

respectively. It is the sharp contrast between Equation (8.35) and Equation (5.16) for Dirac equation (i.e., $\rho^{(D)} = \xi^{(D)\dagger}\xi^{(D)} + \eta^{(D)\dagger}\eta^{(D)}$), that makes Equation (8.15) so unique as shown below.

Let us look at the example of WF for tachyon, Equations (8.16)-(8.18), with $E > 0, p_z > 0$ and $h = -1$. It is allowed just because $|\xi| > |\eta|$ and so $\rho > 0$. Second choice of Equation (8.16) with

$p_z = -p < 0(p > 0), h = +1$ but

$$\xi = \frac{1}{m_s}(-p + E)\eta, |\xi| < |\eta| \tag{8.37}$$

should be fobidden due to its ρ < 0. Another two possible WFs with $\xi \sim \eta \sim \begin{pmatrix} 1 \\ 0 \end{pmatrix}$ have $p_z = p$ and $p_z = -p$ respectively, only the last one with

$p_z = -p < 0(p > 0), h = -1$ is allowed due to its $|\xi| > |\eta|$ and $\rho > 0$.

Let us turn to the solution of Equation (8.31) for antitachyon with $E_c > 0$ by just performing $\mathcal{PT}$ operation on Equation (8.16) yielding:

$$\eta_c \sim \xi_c \sim \exp\left[-i\left(p_z^c z - E_c t\right)\right]\begin{pmatrix} 0 \\ 1 \end{pmatrix} \tag{8.38}$$

Now if $p_z^c = p_c(=|\boldsymbol{p}_c|) > 0$, since

$\sigma_z^c = -\sigma_z, \sigma_z^c \begin{pmatrix} 0 \\ 1 \end{pmatrix} = 1$, so helicity $h_c = 1$. Substitution of Equation (8.38) into Equation (8.33) yields:

$$\eta_c = \frac{1}{m_s}(p_c + E_c)\xi_c, |\eta_c| > |\xi_c| \tag{8.39}$$

which is allowed due to $\rho_c > 0$. Second choice of Equation (8.38) with $p_z^c = -p_c < 0(p_c > 0), h = -1$ but

$$\eta_c = \frac{1}{m_s}(-p_c + E_c)\eta, |\eta_c| < |\xi_c| \tag{8.40}$$

should be forbidden due to its $\rho_c < 0$. In another two possible WFs with $\eta_c \sim \xi_c \sim \begin{pmatrix} 1 \\ 0 \end{pmatrix}$, only that with

$p_z^c = -p_c < 0, h_c = +1$ is allowed due to $\rho_c > 0$.

Hence we see that: The tachyon can only exist in a left-handed (LH) polarized state (with helicity $h = -1$) whereas antitachyon only in a right-handed (RH) polarized state (with $h_c = 1$). We tentatively link this strange feature with that found in neutrinos—only ν_L and $\bar{\nu}_R$ exists in nature whereas ν_R and $\bar{\nu}_L$ are strictly forbidden.

Furthermore, at first sight, although Equation (8.15) certainly has no symmetry under the space inversion $(\boldsymbol{x} \to -\boldsymbol{x}, t \to t)$, it seems to enjoy a pure "timeinversion" $(\boldsymbol{x} \to \boldsymbol{x}, t \to -t)$ symmetry like

$$\begin{cases} \xi(\boldsymbol{x},t) \to \xi(\boldsymbol{x},-t) = \eta_c'(\boldsymbol{x},t) \\ \eta(\boldsymbol{x},t) \to \eta(\boldsymbol{x},-t) = \xi_c'(\boldsymbol{x},t) \end{cases} \tag{8.41}$$

$$\begin{cases} \mathrm{i}\dfrac{\partial}{\partial t}\eta_c' = -\mathrm{i}\boldsymbol{\sigma}\cdot\nabla\eta_c' + m_s\xi_c' \\ \mathrm{i}\dfrac{\partial}{\partial t}\xi_c' = \mathrm{i}\boldsymbol{\sigma}\cdot\nabla\xi_c' - m_s\eta_c' \end{cases} \tag{8.42}$$

We add "′" in the superscript of η_c' to stress that $\eta_c'(\boldsymbol{x},t)$ (being a time reversed WF), though looks like some antitachyon's WF, is obviously different from $\eta_c(\boldsymbol{x},t)$ gained through the $\mathcal{PT}$ inversion, Equation (8.31). Actually, based on Equations (8.29)-(8.31) and (8.41)-(8.42), we have:

$$\begin{cases} \eta_c'(\boldsymbol{x},t) = \eta_c(-\boldsymbol{x},t), \xi_c'(\boldsymbol{x},t) = \xi_c(-\boldsymbol{x},t) \\ \eta_c'(-\boldsymbol{x},t) = \eta_c(\boldsymbol{x},t), \xi_c'(-\boldsymbol{x},t) = \xi_c(\boldsymbol{x},t) \end{cases} \tag{8.43}$$

Interestingly, we cannot find from Equation (8.42) the "physical solution" of $\eta_c'(\boldsymbol{x},t)$ with $|\eta'|_c > |\xi_c'|$ (so $\rho_c > 0$) and $h_c = 1$ (for $\bar{\nu}_R$) simultaneously. Only $\eta_c'(-\boldsymbol{x},t)$ makes physical sense, but it is just $\eta_c(\boldsymbol{x},t)$ like that discussed in Equation (8.39). Notice that the sign change $\boldsymbol{x} \to -\boldsymbol{x}$ in the phase of WF makes a change in the direction of momentum $\boldsymbol{p}_c \to -\boldsymbol{p}_c$. But a WF is always composed of two fields in confrontation, like η_c versus ξ_c here. And the explicit helicity h_c is determined by which one of these two hidden fields being in charge. So the change of $\boldsymbol{x} \to -\boldsymbol{x}$ in these four equalities of Equation (8.43) does reverse the status of η_c versus ξ_c (or η_c' vs ξ_c'), rendering helicity reversed explicitly. The subtlety of tachyon equation, unlike Dirac equation, lies in the fact that only ν_L and $\bar{\nu}_R$ exist whereas ν_R and $\bar{\nu}_L$ are strictly forbidden, i.e., the parity symmetry is violated to maximum. Hence, in strict sense, there is also no physically meaningful WF after the operation of pure "time inversion" on Equation (8.15). We will insist on Equation (8.31) rather than Equation (8.42)—there is only one correct way leading from tachyon to antitachyon via the $\mathcal{PT}$ inversion essentially.

In 2000, Equation (8.25) was first proposed by Tsao Chang and then collaborated with Ni in Ref. [46] (see also [47-52] and the Appendix 9B in Ref. [25]). At first sight, the difference between Equations (8.25) and (8.1) amounts to substituting the mass term βm by $\beta_s m_s$

with $\beta_s = \begin{pmatrix} 0 & I \\ -I & 0 \end{pmatrix}$ being an antihermitian matrix.

Usually, for an equation with nonhermitian Hamiltonian, there is no guarantee for the completeness of its mathematical solutions. In other words, the unitarity of its physical states is at risk. Sometimes, however, a nonhermitian Hamiltonian can be accepted in physics. For example, in the optical model for nuclear physics, an imaginary part of potential, $V = V_0 + \mathrm{i}V_1$, is used to describe the absorption of incident particles successfully. The interesting thing for "tachyonic neutrino" is: Solutions of Equation (8.15) for $E > 0$ $(E_c > 0)$ are coinciding with that for $|\xi| > |\eta|$ $(|\eta_c| > |\xi_c|)$ whereas another would-be solutions with $E > 0$ but $|\xi| < |\eta|$ ($E_c > 0$ but $|\eta_c| < |\xi_c|$) are forbidden, see Equations (8.37) and (8.40). It seems like half of would-be solutions disappear automatically. Equivalently, from physical point of view, only half of states with $\rho > 0$ or $\rho_c > 0$ are allowed in nature whereas another half with $\rho < 0$ or $\rho_c < 0$ are not. Hence one unique feature of "tachyon" equation, like Equation (8.15) or (8.26), lies in its strange realization of unitarity violation that half of would-be states (being tentatively identified with ν_R and $\overline{\nu}_L$) are absolutely forbidden whereas another half (ν_L and $\overline{\nu}_R$) are stabilized. The permanently longitudinal polarization property of neutrino and antineutrino like that analysed above was first predicted by Lee and Yang in 1957 [3-5] and had been verified by GGS experiment in 1958 [53]. Further discussion on this topic is currently in preparation.

9. ANTIGRAVITY BETWEEN MATTER AND ANTIMATTER

In hindsight, there are two Lorentz invariants in the kinematics of SR:

$$c^2(t_1-t_2)^2-(\boldsymbol{x}_1-\boldsymbol{x}_2)^2$$
$$=c^2(t_1'-t_2')^2-(\boldsymbol{x}_1'-\boldsymbol{x}_2')^2=\text{const} \tag{9.1}$$

$$E^2-c^2\boldsymbol{p}^2=E'^2-c^2\boldsymbol{p}'^2=m^2c^4 \tag{9.2}$$

It seems quite clear that Equation (9.1) is invariant under the space-time inversion $(\boldsymbol{x}\to-\boldsymbol{x},t\to-t)$ and Equation (9.2) remains invariant under the mass inversion $(m\to-m)$. We believe that these two discrete symmetries are deeply rooted at the SR's dynamics via its combination with QM and developing into RQM and QFT—the particle and its antiparticle are treated on equal footing and linked by the symmetry $\mathcal{PT}=\mathcal{C}$ essentially. Hence we can perform a mass inversion on Equation (9.2) in each of two inertial frames with arbitrary relative velocity $\boldsymbol{v}$ in the sense of $m\to-m_c=-m,E\to-E_c,\boldsymbol{p}\to-\boldsymbol{p}_c$, yielding:

$$E_c^2-c^2\boldsymbol{p}_c^2=E_c'^2-c^2\boldsymbol{p}_c'^2=m_c^2c^4=m^2c^4 \tag{9.3}$$

The invariance of Equation (9.2) under mass inversion as a whole reflects the experimental fact that particle and antiparticle are equally existing in nature even at the level of classical physics.

Example: The motion equation for a charged particle (say, electron with charge $q=-e<0$) in the external electric and magnetic fields, $\boldsymbol{E}$ and $\boldsymbol{B}$, is given by the Lorentz formula:

$$m\boldsymbol{a}=q\left(\boldsymbol{E}+\frac{1}{c}\boldsymbol{v}\times\boldsymbol{B}\right) \tag{9.4}$$

Then the operation of either $q\to q_c=-q$ or $m\to-m_c=-m$ on Equation (9.4) will realize the transformation from particle into its antiparticle (say, positron with charge $q_c=-q=e>0$) with the acceleration change from $\boldsymbol{a}\to\boldsymbol{a}_c=-\boldsymbol{a}$ as

$$ma_c = -q\left(E + \frac{1}{c}v \times B\right) \tag{9.5}$$

Based on what we learn from RQM (Sections III-V) as well as Equations (9.1)-(9.5), we may conjecture that for a classical theory being capable of treating matter and antimatter on an equal footing, it must be invariant under a mass inversion $m \to -m_c = -m$.

Notice that, however, Equation (9.4) (Equation (9.5)) is only valid for particle (antiparticle) moving at low speed, it must be modified to adapt to high-speed cases through the invariance of continuous Lorentz transformation. So we need "double checks" for testing a classical theory being really "relativistic" or not.

Let us restudy the theory of general relativity (GR). In a $(-,+,+,+)$ metric, the Einstein field equation (EFE) reads (see, e.g. , Refs. [54-56]) $(c=1)$,

$$G_{\mu\nu} \equiv R_{\mu\nu} - \frac{1}{2}g_{\mu\nu}R = -8\pi G T_{\mu\nu} \tag{9.6}$$

Of course, Equation (9.6) is covariant with respect to the Lorentz transformation. But could it withstand the test of mass inversion?

On the LHS of Equation (9.6), the Einstein tensor $G_{\mu\nu}$ contains no any mass and no charge as well. But on the RHS, the energy-momentum current density tensor $T_{\mu\nu}$ is proportional to particle's mass m and so changes its sign under an operation of $m \to -m$. Hence as a whole, Equation (9.6) cannot remain invariant under the mass inversion. The reason seems rather clear that antimatter was not taking into account when GR was established in 1915. To modify EFE such that it can preserve the invariance of mass inversion, in 2004, one of us (Ni) proposed to add another term with $T^c_{\mu\nu}$ for antimatter, yielding [27]

$$R_{\mu\nu} - \frac{1}{2}g_{\mu\nu}R = -8\pi G\left(T_{\mu\nu} - T^c_{\mu\nu}\right) \tag{9.7}$$

which remains invariant under a mass inversion since:

$$T_{\mu\nu} \to -T^c_{\mu\nu}, T^c_{\mu\nu} \to -T_{\mu\nu} \, (m \to -m) \tag{9.8}$$

In a weak-field (or the post-Newtonian) approximation, this modified EFE, MEFE, Equation (9.7), will lead to modified Newton gravitational law as

$$F_{\text{grav}}(r) = \mp G \frac{mm'}{r^2} \tag{9.9}$$

where the "$-$" sign means attractive force between m and m' being both matter or antimatter whereas the "+" sign means repulsive force between m and m' (both positive) if one of them is antimatter.

If we define the "gravitational mass" for matter and antimatter separately

$$m_{\text{grav}} = \begin{cases} m > 0, & (\text{matter}) \\ -m_c = -m < 0, & (\text{antimatter}) \end{cases} \tag{9.10}$$

Then Equation (9.9) can be recast into one equation

$$F_{\text{grav}}(r) = -G \frac{m_{\text{grav}} m'_{\text{grav}}}{r^2} \tag{9.11}$$

which bears a close resemblance to the Coulomb law in classical electrodynamics (CED)

$$F_{\text{Coul}}(r) = \frac{qq'}{r^2} \tag{9.12}$$

In 1986, within the framework of classical field theory (CFT) plus some assumptions, Jagannathan and Singh derived the potential energy of two static point sources as [57]

$$U(r) = (-1)^{n+1} ee' \times (\text{a positive number}) \times \frac{e^{-\mu r}}{r} \tag{9.13}$$

where n and μ are spin and mass of the mediating field, e is the "charge" of the source. For CED, $n = 1$ whereas $n = 2$ for gravitational field

($\mu \to 0$ in both cases). So Equation (9.13) is in conformity with Equations (9.11) and (9.12) for the case of "like sources" (with $ee' > 0$) [57], where the case for "unlike sources" $(ee' < 0)$ hadn't been discussed. Here Equation (9.11) has been generalized to the case for "unlike sources", but at a price that the "equivalence principle" in GR ceases to be valid when matter and antimatter coexist as shown by Equation (9.10).

In 2011, the antigravity between matter and antimatter was also claimed by Villata in Ref.[58], where the argument seems different from that explained above. But theory is theory, only fact will have the final say. So we are anxiously waiting for the outcome from the AEGIS experiment [59] (at CERN), which is designed to compare the Earth gravitational acceleration on hydrogen and antihydrogen atoms.

10. SUMMARY

1) Being the combination of SR and QM, RQM is capable of dealing with particle and antiparticle on an equal footing. As long as we admit that the antiparticle's momentum and energy operators should be $\hat{p}_c = i\hbar\nabla$ and $\hat{E}_c = -i\hbar\frac{\partial}{\partial t}$ versus $\hat{p} = -i\hbar\nabla$ and $\hat{E} = i\hbar\frac{\partial}{\partial t}$ for particle, it can be proved that the "negative-energy" WF ψ of particle corresponds to a "positive-energy" WF ψ_c of antiparticle precisely.

2) In general, an equation in RQM always has a discrete symmetry $\mathcal{PT} = \mathcal{C}$ which shows up as a transformation between a particle's WF ψ and its antiparticle's WF ψ_c: $\psi(x,t) \leftrightarrows \psi_c(x,t)$. For a free particle, it simply means $\psi(-x,-t) = \psi_c(x,t)$. This is in conformity with the "strong reflection" in QFT invented by Pauli and Lüders, showing that the intrinsic property of a particle cannot be detached from the space-time.

3) Following Feshbach-Villars' deep insight, we are able to divide each and every WF ψ in RQM into two parts, $\psi = \phi + \chi$. Then the above symmetry is further rigorously expressed by an invariance of motion

equation in RQM through the transformations $\phi \leftrightarrows \chi_c$ and $\chi \leftrightarrows \phi_c$ under either the space-time inversion

$(x \to -x, t \to -t)$ or a mass inversion $(m \to -m)$. Since $|\phi| > |\chi|$ in ψ whereas $|\chi_c| > |\phi_c|$ in ψ_c, we may name ϕ as the (dominant) hidden particle field in ψ while χ the (subordinate) hidden antiparticle field in ψ. In this way, both the "probability density" ρ for a particle and ρ_c for an antiparticle can be proved to be positive definite. Now we may say that the RQM is ensured to be self-consistent and can be regarded as a sound basis for QFT.

4) All kinematical effects in SR can be ascribed to the enhancement of the magnitude of χ field in a particle's WF accompanying with the increase of particle's velocity.

5) As proved for Dirac particle with spin, the helicity of a particle is just opposite to that of its antiparticle under a space-time (or mass) inversion. Therefore, the experimental tests for the CPT invariance should include not only the equal mass and lifetime of particle versus antiparticle, but also the following fact: A particle and its antiparticle with opposite helicities must coexist in nature with no exception. A prominent example is the neutrino —

A neutrino ν_L (antineutrino $\bar{\nu}_R$) is permanently lefthanded (right-handed) polarized whereas the fact that no ν_R exists in nature must means no $\bar{\nu}_L$ as well (as verified by the GGS experiment [53]). See also Section VII.

6. Based on the invariance of space-time inversion or mass inversion (at the level of RQM) and the latter's generalization to the classical physics, we tentatively discuss some interesting problems in today's physics, including the prediction of antigravity between matter and antimatter, as well as the reason why we believe neutrinos are likely the tachyons.

11. ACKNOWLEDGEMENTS

We thank E. Bodegom, T. Chang, Y. X. Chen, T. P. Cheng, X. X. Dai, G. Tananbaum, V. Dvoeglazov, Y. Q. Gu, F. Han, J. Jiao, A. Kellerbauer, T. C. Kerrigan, A. Khalil, R. Konenkamp, D. X. Kong, J. S. Leung, P. T. Leung, Q. G. Lin, S. Y. Lou, D. Lu, Z. Q. Ma, D. Mitchell, E. J. Sanchez, Z. Y. Shen, Z. Q. Shi, P. Smejtek, X. T. Song, R. K. Su, G. Tananbaum Y. S. Wang, Z. M. Xu, X. Xue, J.

Yan, F. J. Yang, J. F. Yang, R. H. Yu, Y. D. Zhang and W. M. Zhou for encouragement, collaborations and helpful discussions.

REFERENCES

1. A. Apostolakis, et al., (CPLEAR Collaboration) Physics Letters B, Vol. 422, 1998, pp. 339-348.
2. H. Feshbach and F. Villars, Review of Modern Physics, Vol. 30, 1958, pp. 24-45.
3. T. D. Lee and C. N. Yang, Physical Review, Vol. 104, 1956, pp. 254-258.
4. T. D. Lee and C. N. Yang, ibid, Vol. 105, 1957, pp. 1671-1675.
5. T. D. Lee, R. Oehme and C. N. Yang, ibid, Vol. 106, 1957, pp. 340-345.
6. C. S. Wu, E. Ambler, R. W. Hayward, D. D. Hoppes and R. P. Hudson, Physical Review, Vol. 105, 1957, pp. 1413-1415.
7. J. H. Christensen, J. W. Cronin, V. L. Fitch and R. Turlay, Physical Review Letters, Vol. 13, 1964, pp. 138-140.
8. K. R. Schubert, B. Wolff, J.-M. Gaillard, M. R. Jane, T. J. Ratcliffe and J.-P. Repellin, Physics Letters B, Vol. 31, 1970, pp. 662-665.
9. J. Beringer, et al., (Particle Data Group) Physical Review D, Vol. 86, 2012, Article ID: 010001.
10. G. Lüders, Kgl. Danske Vidensk. Selsk. Mat.-Fys. Medd., Vol. 28, 1954.
11. G. Lüders, Annals of Physics (New York), Vol. 2, 1957, pp. 1-15.
12. W. Pauli, "Exclusion Principle, Lorentz Group and Reflection of Space-Time and Charge," In: W. Pauli, L. Rosenfeld and V. Weisskopf, Eds., Niels Bohr and the Development of Physics, McGraw-Hill, New York, 1955, pp. 30-51.
13. T. D. Lee and C. S. Wu, Annual Review of Nuclear Science, Vol. 15, 1965, pp. 381-476.
14. A. Einstein, B. Podolsky and N. Rosen, Physical Review, Vol. 47, 1935, pp. 777-780.

15. D. Bohm, "Quantum Theory," Prentice Hall, Upper Saddle River, 1956.

16. J. S. Bell, Physics, Vol. 1, 1964, pp. 195-200.

17. H. Guan, "Basic Concepts in Quantum Mechanics," High Education Press, Beijing, 1990.

18. G. J. Ni, H. Guan, W. M. Zhou and J. Yan, Chinese Physics Letters, Vol. 17, 2000, pp. 393-395.

19. O. Nachtmann, "Elementary Particle Physics: Concepts and Phenomena," Springer-Verlag, Berlin, 1990.

20. W. Greiner and B. Müller, "Gauge Theory of Weak Interactions," Springer-Verlag, Berlin, 1993.

21. E. J. Konopinski and H. M. Mahmaud, Physical Review, Vol. 92, 1953, pp. 1045-1049.

22. G. J. Ni, Journal of Fudan University (Natural Science), No. 3-4, 1974, pp. 125-134.

23. G. J. Ni and S. Q. Chen, Journal of Fudan University (Natural Science), Vol. 35, 1996, pp. 325-334.

24. G. J. Ni and S. Q. Chen, "Relation between Space-Time Inversion and Particle-Antiparticle Symmetry and the Microscopic Essence of Special Relativity," In: V. Dvoeglazov, Ed., Photon and Poincare Group, NOVA Science Publisher, New York, 1999, pp. 145-169.

25. G. J. Ni and S. Q. Chen, "Advanced Quantum Mechanics," Rinton Press, New Jersy, 2002.

26. G. J. Ni, Progress in Physics, Vol. 23, 2003, pp. 484-503.

27. G. J. Ni, "A New Insight into the Negative-Mass Paradox of Gravity and the Accelerating Universe," In: V. V. Dvoeglazov and A. A. Espinoza Garrido, Eds., Relativity, Gravitation, Cosmology, NOVA Science Publisher, New York, 2004, pp. 123-136.

28. G. J. Ni, J. J. Xu and S. Y. Lou, Chinese Physics B, Vol. 20, 2011, Article ID: 020302.

29. J. J. Sakurai, "Advanced Quantum Mechanics," Addison-Wesley Publishing Company, Boston, 1978.

30. J. J. Sakurai, "Modern Quantum Mechanics," John Wiley & Sons, Inc., NewYork, 1994.

31. J. D. Bjorken and S. D. Drell, "Relativistic Quantum Mechanics," McGraw-Hill, New York, 1964,

32. J. D. Bjorken and S. D. Drell, "Relativistic Quantum Fields," McGraw-Hill, New York, 1965.

33. L. B. Okun, Physics Today, Vol. 42, 1989, pp. 31-36.

34. G. Lochak, "De Broglie's Initial Conception of De Broglie Waves," In: S. Diner, D. Fargue, G. Lochak and F. Selleri, Eds., The Wave-Particle Dualism, D. Reidal Publishing Company, Dordrecht, 1984, pp. 1-25.

35. G. J. Ni, W. M. Zhou and J. Yan, "Comparison among Klein-Gordon Equation, Dirac Equation and Relativistic Schrödinger Equation," In: A. E. Chubykalo, V. V. Dvoeglazov, D. J. Ernst, V. G. Kadyshevsky and Y. S. Kim, Eds., Lorentz Group, CPT and Neutrinos, World Scientific, London, 2000, pp. 68-81.

36. M. E. Peskin and D. V. Schroeder, "An Introdution to Quantum Field Theory," Addison-Wesley Publishing Company, Boston, 1995.

37. M. Jacob and G. C. Wicks, Annals of Physics (New York), Vol. 7, 1959, pp. 404-428.

38. S. Weinberg, Physical Review Letters, Vol. 19, 1967, pp. 1264-1266.

39. Z. Q. Shi and G. J. Ni, Chinese Physics Letters, Vol. 19, 2002, pp. 1427-1429.

40. Z. Q. Shi and G. J. Ni, Annales de la Fondation Louis de Bloglie, Vol. 29, 2004, pp. 1057-1066.

41. Z. Q. Shi and G. J. Ni, Handronic Journal, Vol. 29, 2006, pp. 401-407.

42. Z. Q. Shi and G. J. Ni, "Frontiers in Horizons in World Physics," Nova Science, Marselle, 2008, pp. 53-65.

43. Z. Q. Shi and G. J. Ni, Modern Physics Letters A, Vol. 26, 2011, pp. 987-998.

44. A. Cho, Science, Vol. 326, 2009, pp. 1342-1343.

45. L. H. Ryder, "Quantum Field Theory," Cambridge University Press, Cambridge, 1996.

46. T. Chang and G. J. Ni, "An Explanation of Possible Negative Mass-Square of Neutrinos," FIZIKA B (Zagreb), Vol. 11, 2002, pp. 49-56. arXiv.org:hep-ph/0009291

47. G. J. Ni and T. Chang, Journal of Shaanxi Normal University (Natural Science), Vol. 30, No. 3, 2002, pp. 32-39.

48. G. J. Ni, Journal of Shaanxi Normal University (Natural Science), Vol. 29, No. 1, 2001, pp. 1-5.

49. G. J. Ni, Journal of Shaanxi Normal University (Natural Science), Vol. 30, No. 4, 2002, pp. 1-6.

50. G. J. Ni, "A Minimal Three-Flavor Model for Neutrino Oscillation Based on Superluminal Property," In: V. V. Dvoeglazov and A. A. Espinoza, Eds., Relativity, Gravitation, Cosmology, NOVA Science Publisher, New York, 2004, pp. 137-148.

51. G. J. Ni, "Principle of Relativity in Physics and in Epistemology," In: V. Dvoeglazov, Ed., Relativity, Gravitation, Cosmology: New Development, NOVA Science Publisher, New York, 2010, pp. 237-252.

52. G. J. Ni, "Cosmic Ray Spectrum and Tachyonic Neutrino," In: V. V. Dvoeglazov and A. A. Espinoza, Eds., Relativity, Gravitation, Cosmology: New Development, NOVA Science Publisher, New York, 2010, pp. 253-265.

53. M. Goldhaber, L. Grodgins and A. W. Sunyar, Physical Review, Vol. 109, 1958, pp. 1015-1017.

54. S. Weinberg, "Gravitation and Cosmology," John Wiley, New York, 1972.

55. Z. M. Xu and X. J. Wu, "General Relativity and Contemporary Cosmology," Press of Nanjing Normal University, Nanjing, 1999.

56. T. P. Cheng, "Relativity, Gravitation and Cosmology", 2nd Edition, Oxford University Press, Oxford, 2010.

57. K. Jagannathan and L. P. S. Singh, Physical Review D, Vol. 33, 1986, pp. 2475-2477.

58. M. Villata, Europhysics Letters, Vol. 94, 2011, pp. 1-6.

59. A. Kellerbauer, et al., Nuclear Instruments and Methods in Physics Research Section B, Vol. 266, 2008, pp. 351- 356.

60. O. Klein, Zeitschrift für Physik, Vol. 53, 1929, pp. 157-165.

61. W. Greiner, "Relativistic Quantum Mechanics," SpringerVerlag, Berlin, 1990.

62. W. Greiner, B. Müller and J. Rafelski, "Quantum Electrodynamics of Strong Fields," Springer-Verlag, Berlin, 1985.

6
CHAPTER

CALCULATION OF THE APPROXIMATE ENERGY OF GROUND AND EXCITED STATIONARY STATES IN QUANTUM MECHANICS USING DELTA METHOD

Farrin Payandeh, Touraj Mohammadpour

Department of Physics, Payame Noor University (PNU), Tehran, Iran

ABSTRACT

In this paper, pursuing a new advised method called Delta method which is basically similar to variational method, we find the ground and excited states, according to a typical quantum Hamiltonian. Moreover, applying this method, the upper bound values for the eigenenergies of the so- called ground and excited states are estimated. We will show that this new method, is as beneficial as the traditional variational method which is common in deriving eigenenergies of some of the quantum Hamiltonians. This method helps physics students to broaden their knowledge about the possible mathematical ways; they can use to obtain eigenenergies of some quantum Hamiltonians. The advantage of Delta method to variational method is in its simplicity and reduction of the calculation procedures.

Keywords: Quantum Mechanics, Eigenenergies, Alternative Methods, Delta Method

1. INTRODUCTION

Most problems encountered in quantum mechanics cannot be solved exactly. Exact solutions of the Schrodinger equation exist only for a few idealized systems. So, in order to solve general problems, one must resort to approximation methods. Up to now, a variety of such methods have been developed, and each has its own area of applicability. There are many methods for solving Schrodinger equation, i.e. perturbation theory [1] , the variational method [1] , and the WKB method [1] , Supersymmetry quantum mechanics [2] - [6] , Nikivorov-Uvarov method [7] - [9] , Romanovski polynomials in quantum mechanics [10] - [12] , etc. [12] - [23] .

Three conventional approximation methods for studying the stationary states corresponding to time-indepen- dent Hamiltonians, are: perturbation theory, the variational method, and the WKB method. Perturbation theory is based on the assumption that the problem we wish to solve is, in some sense, only slightly different from a problem that can be solved exactly. In the case where the deviation between the two problems is small, perturbation theory is suitable for calculating the contribution associated with this deviation; this contribution is then added as a correction to the energy and the wave function of the exactly solvable Hamiltonians. So perturbation theory builds on the known exact solutions to obtain approximate solutions.

But, about those systems whose Hamiltonians cannot be reduced to an exactly solvable part plus a small correction, the variational method or the WKB approximation are considered. The variational method is particularly useful in estimating the energy eigenvalues of the ground state and the first few excited states of a system for which one has only a qualitative idea about the form of the wave function. The WKB method is useful for finding the energy eigenvalues and wave functions of systems for which the classical limit is valid. Unlike perturbation theory, the variational and WKB methods do not require the existence of a closely related Hamiltonian that can be solved exactly [1] .

The application of the approximation methods to the study of stationary states consists of finding the energy eigenvalues E_n and the eigenfunctions $|\psi_n\rangle$ of a time-independent Hamiltonian $\hat{H}$ that does not have exact solutions:

$$\hat{H}|\psi_n\rangle = E_n|\psi_n\rangle$$

Depending on the structure of $\hat{H}$, we can use any of the three methods mentioned above to find the approximate solutions to this eigenvalue problem.

In this paper, we will use a new approximation method called Delta method for finding the ground and excited energy state of stationary states. This method, with a difference in the way of calculation, is somehow similar to the variational method at the beginning. Like the variational method, we first find the time-independent Hamiltonian $\hat{H}$ that does not have exact solution, using a supposed trial function and calculating the value of energy E_n in terms of a parameter α. Then, we will estimate the upper bound values for the eigenenergies of ground and excited states applying the Delta conditions $\Delta \geq 0$ and $D \leq 0$ on the second and third order equations, respectively. The difference between Delta and variational methods is that in variational method the derivative of E to α is calculated and the solution for α is obtained with affecting the conditions on E. But, in Delta method, we obtain the physical solutions with writing α in terms of E in the form of a two or three order equation and applying the mathematical Delta conditions. Moreover, in variational method, after calculating the derivative of E to α and equating it with zero to find the α which minimizes $E(\alpha)$, i.e. in Hydrogen atom, sometimes one of the solutions for α is infinite which is an unphysical solution, since it leads to a zero value for the ground or excited state energy. While, in delta method, only the physical solution is attained.

It should be noted that Delta method could be applied to all of the problems to be solved through variational method and exactly give the same result. However, the advantage of Delta method is first in its simplicity, and then in reduction of the calculation procedures.

In Section 2, we will have a review on the variational method [1] . In Section 3, we will explain Delta method and in Section 4, we will show the applicability and simplicity of Delta method with some examples.

2. VARIATIONAL METHOD

There exist systems whose Hamiltonians are known, but they cannot be solved exactly or by a perturbative treatment. That is, there is no closely related Hamiltonian that can be solved exactly or approximately by perturbation theory because the first order is not sufficiently accurate. One of the approximation methods that are suitable for solving such problems is the variational method, which is also called the Rayleigh-Ritz method. This method does not require knowledge of simpler Hamiltonians that can be solved exactly. The variational method is useful for determining upper bound values for the eigenenergies of a system whose Hamiltonian is known whereas its eigenvalues and eigenstates are not known. It is particularly useful for determining the ground state. It becomes quite cumbersome to determine the energy levels of the excited states.

In the context of the variational method, one does not attempt to solve the eigenvalue problem:

$$\hat{H}|\psi_n\rangle = E_n|\psi_n\rangle \tag{1}$$

But rather one uses a variational scheme to find the approximate eigenenergies and eigenfunctions from the variational equation:

$$\delta E(\psi) = 0 \tag{2}$$

where $E(\psi)$ is the expectation value of the energy in the state $|\psi\rangle$:

$$E(\psi) = \frac{\langle\psi|\hat{H}|\psi\rangle}{\langle\psi|\psi\rangle} \tag{3}$$

If $|\psi\rangle$ depends on a parameter α, $E(\psi)$ will also depend on α. The variational ansatz (2) enables us to vary α so as to minimize $E(\psi)$. The minimum value of $E(\psi)$ provides an upper limit approximation for

the true energy of the system. The variational method is particularly useful for determining the ground state energy and its eigenstate without explicitly solving the Schrodinger equation. Not that for any (arbitrary) trial function $|\psi\rangle$ we choose, the energy E as given by (3) is always larger than the exact energy E_0:

$$E = \frac{\langle\psi|\hat{H}|\psi\rangle}{\langle\psi|\psi\rangle} \geq E_0 \tag{4}$$

The equality condition occurs only when $|\psi\rangle$ is proportional to the true ground state $|\psi_0\rangle$. Because, expanding the trial function $|\psi\rangle$ in terms of the exact eigenstates of $\hat{H}$:

$$|\psi\rangle = \sum_n a_n |\phi_n\rangle \tag{5}$$

with

$$\hat{H}|\phi_n\rangle = E_n |\phi_n\rangle \tag{6}$$

and since $E_0 \geq E_n$ for nondegenerate one-dimensional bound systems, we have:

$$E = \frac{\langle\psi|\hat{H}|\psi\rangle}{\langle\psi|\psi\rangle} = \frac{\sum_n |a_n|^2 E_n}{\sum_n |a_n|^2} \geq \frac{E_0 \sum_n |a_n|^2}{\sum_n |a_n|^2} = E_0 \tag{7}$$

which proves (4).

To calculate the ground state energy, we need to carry out the following four steps:

- First, based on physical intuition, make an educated guess of a trial function that takes into account all the physical properties of the ground state (symmetries, number of nodes, smoothness, behavior at infinity, etc.). For the properties we are not sure about, they can be included in the trial function adjustable parameters $\alpha_1, \alpha_2, \cdots$ (i.e., $|\psi_0\rangle = |\psi_0(\alpha_1, \alpha_2, \cdots)\rangle$) which will account for the various possibilities of these unknown properties.
- Second, using (3), calculate the energy; this yields an expression which depends on the parameters $\alpha_1, \alpha_2, \cdots$:

$$E_0(\alpha_1, \alpha_2, \cdots) = \frac{\langle\psi_0(\alpha_1, \alpha_2, \cdots)|\hat{H}|\psi_0(\alpha_1, \alpha_2, \cdots)\rangle}{\langle\psi_0(\alpha_1, \alpha_2, \cdots)|\psi_0(\alpha_1, \alpha_2, \cdots)\rangle} \tag{8}$$

In most cases $|\psi_0(\alpha_1, \alpha_2, \cdots)\rangle$ will be assumed to be normalized; hence the denominator of this expression is equal to 1.

- Third, using (8) search for the minimum of $E_0(\alpha_1, \alpha_2, \cdots)$ by varying the adjustable parameters α_i until E_0 is minimized. That is, minimize $E(\alpha_1, \alpha_2, \cdots)$ with respect to $\alpha_1, \alpha_2, \cdots$:

$$\frac{\partial E_0(\alpha_1, \alpha_2, \cdots)}{\partial \alpha_i} = \frac{\partial}{\partial \alpha_i} \frac{\langle \psi_0(\alpha_1, \alpha_2, \cdots) | \hat{H} | \psi_0(\alpha_1, \alpha_2, \cdots) \rangle}{\langle \psi_0(\alpha_1, \alpha_2, \cdots) | \psi_0(\alpha_1, \alpha_2, \cdots) \rangle} = 0 \quad (9)$$

with $i = 1, 2, \cdots$. This gives the values of $(\alpha_{1_0}, \alpha_{2_0}, \cdots)$ that minimize E_0.

- Fourth, substitute these values of $(\alpha_{1_0}, \alpha_{2_0}, \cdots)$ into (8) to obtain the approximate value of the energy. The value of $E_0(\alpha_{1_0}, \alpha_{2_0}, \cdots)$ thus obtained, provides an upper bound for the exact ground state energy E_0. The exact ground state eigenstate $|\phi_0\rangle$ will then be approximated by the state $|\psi_0(\alpha_{1_0}, \alpha_{2_0}, \cdots)\rangle$.

About the energies of the excited states, it should be said that the variational method can also be used to find the approximate values for the energies of the first few exited states. For instance, to find the energy and eigenstate of the first excited state that will approximate E_1 and $|\phi_1\rangle$, we need to choose a trial function $|\psi_1\rangle$ that must be orthogonal to $|\psi_0\rangle$:

$$\langle \psi_1 | \phi_0 \rangle = 0 \quad (10)$$

Then proceed as we did in the case of the ground state. That is, solve the variational Equation (2) for $|\psi_1\rangle$:

$$\frac{\partial}{\partial \alpha_i} \frac{\langle \psi_1(\alpha_1, \alpha_2, \cdots) | \hat{H} | \psi_1(\alpha_1, \alpha_2, \cdots) \rangle}{\langle \psi_1(\alpha_1, \alpha_2, \cdots) | \psi_1(\alpha_1, \alpha_2, \cdots) \rangle} = 0; \quad (i = 1, 2, \cdots) \quad (11)$$

Similarly, to evaluate the second excited state, we solve (2) for $|\psi_2\rangle$ and take into account the following two conditions:

$$\langle \psi_2 | \psi_0 \rangle = 0, \qquad \langle \psi_2 | \psi_1 \rangle = 0 \tag{12}$$

These conditions can be included in the variational problem by means of Lagrange multipliers, that is, by means of a constrained variational principle.

In this way, we can in principle evaluate any other excited state. However, the variational procedure becomes increasingly complicated as we deal with higher excited states. As a result, the method is mainly used to determine the ground state.

3. DELTA METHOD

In this section, we obtain the expectation value of energy in a supposed state $|\psi\rangle$, like the variational method, $|\psi\rangle$ depends on a parameter α, and finally $E(\psi)$ will depend on α, too. Then, instead of minimizing $E(\psi)$ leading to a minimum value of $E(\psi)$ as an upper approximate limit for the real energy of system, we use a new method in which we write the expectation value of energy in a supposed state $|\psi\rangle$ in the form of a second order equation in terms of α, i.e. $a\alpha^2 + b\alpha + c = 0$, and using the method of finding the solution of a second order equation, we calculate $\Delta = \sqrt{b^2 - 4ac}$ and then, write α in terms of $E(\psi)$, so that we have:

$$\alpha_{1,2} = \frac{-b \pm \sqrt{b^2 - 4ac}}{2a} \tag{13}$$

where a, b, c are constants and the term $b^2 - 4ac$ will be a function of E. For having a solution for α in terms of E, we should have:

$$\Delta \geq 0 \Rightarrow b^2 - 4ac \geq 0 \tag{14}$$

Then, using this inequality, just like the variational method, an upper limit for energy is obtained that leads to finding the approximate value of ground or excited state energy.

In some of the problems, the expectation value of energy in terms of α leads to a third order equation in terms of E in the form of $a\alpha^3 + b\alpha^2 + c\alpha + d = 0$ using which by writing the required conditions for having real roots of α in terms of E, one can find the desired solutions, i.e. the ground or excited state energy. Writing the required conditions for solving third order equation $ax^3 + bx^2 + cx + d = 0$, we first write the parameters f, q, D as the following:

$$P = \frac{1}{3}\left(\frac{3c}{a} - \frac{b^2}{a^2}\right) \tag{15}$$

$$q = \frac{1}{27}\left(\frac{2b^3}{a^3} - \frac{9bc}{a^2} + \frac{27d}{a}\right) \tag{16}$$

$$D = \left(\frac{p}{3}\right)^3 + \left(\frac{q}{2}\right)^2 \tag{17}$$

Now, according the sign of D, the solutions for the third order equation are obtained as:

$$\text{if } D \leq 0 : \phi = \arccos\left[-\frac{q}{2}\left(\frac{|p|}{3}\right)^{-\frac{3}{2}}\right]$$

$$y_1 = 2\left(\frac{|p|}{3}\right)^{\frac{1}{2}} \cos\frac{\phi}{3} \tag{18}$$

$$y_{2,3} = -2\left(\frac{|p|}{3}\right)^{\frac{1}{2}} \cos\frac{\phi \pm \pi}{3} \tag{19}$$

Then, three distinct real solutions are obtained for α in terms of E. Moreover:

$$\text{if } D>0; \quad u=\left(-\frac{q}{2}+D^{\frac{1}{2}}\right)^{\frac{1}{3}}, \quad v=\left(-\frac{q}{2}-D^{\frac{1}{2}}\right)^{\frac{1}{3}}$$

$$y_1 = u+v, \tag{20}$$

$$y_{2,3} = -\frac{(u+v)}{2} \pm i\left(\frac{u-v}{2}\right)3^{\frac{1}{2}} \tag{21}$$

then, one real solution and two complex solutions are obtained [24] . Of course, since in physical problems α is a real scale factor, so without the need for calculating α in terms of E, it is sufficient to apply the Delta conditions $\Delta \geq 0$ and $D \leq 0$ on the second and third order equations respectively, and estimate the ground or excited state energies. It should be noted that the reason for using the condition $D \leq 0$ is to ensure the real solutions. This method can be applied for higher order equations in terms of α , too.

In the next section, we will consider a few examples of quantum mechanics and estimate their upper limit of the ground or excited state energies, using Delta method. We remark that this method can be applied as another approach to approximate solutions for all the problems of quantum mechanics to be solved through the variational method. In the case of selecting a suitable trial function, Delta method can be applied to a complex problem such as Poschl-Teller, Rosen-Morse, or another shape invariant potential in quantum mechanics, too. However, the main point is that proposing and guessing the suitable trial function which is based on physical intuition, is very important. For example, it can be shown that with some proposal trial wave functions and applying some simplification, exactly the same result could be achieved for Poschl-Teller and Rosen-Morse potentials through both variational and Delta methods. Of course, it should be noted that up to now, a variety of approximation methods have been developed, and each has its own area of applicability. However, the main purpose of introducing Delta method is to show that this new method, is as beneficial as the traditional variational method which is common in deriving eigenenergies of some of the quantum Hamiltonians, and to help physics students to broaden their knowledge about the possible mathematical ways to obtain eigenenergies of some quantum Hamiltonians with a simpler method than the conventional variational method.

4. EXAMPLES

In this section, using Delta method, we estimate the upper limit of the ground or excited state energies for a few examples of quantum mechanics that are obtained from the variational method, before.

Exp. 1

Calculating the ground and excited state energies of a one-dimensional harmonic oscillator using Delta method.

Choosing the trial function for ground state in the form of $\psi_0(x,\alpha) = Ae^{-\alpha x^2}$ and writing the expectation value of E₀ in terms of α :

$$E_0(\alpha) = \langle\psi_0|H|\psi_0\rangle = A^2\int_{-\infty}^{+\infty} e^{-\alpha x^2}\left(-\frac{\hbar^2}{2m}\frac{d^2}{dx^2} + \frac{1}{2}m\omega^2x^2\right)e^{-\alpha x^2}dx$$

Neglecting the calculations, we get:

$$E_0(\alpha) = \frac{\hbar^2}{2m}\alpha + \frac{m\omega^2}{8\alpha}$$

In Delta method there is no need for obtaining the derivative of $E_0(\alpha)$ to α .

It is just sufficient to form the second order equation for α in terms of E:

$$E_0(\alpha) = \frac{\hbar^2 4\alpha^2 + m^2\omega^2}{8m\alpha}$$

Using the Delta method, the solutions for α will be:

$$\alpha_{1,2} = \frac{8mE_0(\alpha) \pm \sqrt{(-8mE_0(\alpha))^2 - 4\alpha 4\hbar^2 m^2\omega^2}}{2\alpha 4\hbar^2}$$

$$(8mE_0(\alpha))^2 \geq 16\hbar^2m^2\omega^2 \Rightarrow 8mE_0(\alpha) \geq 4\hbar m\omega \Rightarrow E_0(\alpha) \geq \frac{\hbar\omega}{2}$$

As it can be seen, $\frac{\hbar\omega}{2}$ is an upper limit for $E_0(\alpha)$ which is the value of the ground state energy. For obtaining the excited state energy, writing the trial function in the form of $\psi_1(x,\alpha) = Bxe^{-E_0(\alpha)x^2}$, we have:

$$E_1(\alpha)=\langle\psi_1(\alpha)|H|\psi_1(\alpha)\rangle=B^2\int_{-\infty}^{+\infty}xe^{-\alpha x^2}\left[-\frac{\hbar^2}{2m}\frac{d^2}{dx^2}+\frac{1}{2}m\omega^2x^2\right]xe^{-\alpha x^2}dx$$

$$=\frac{3\hbar^2}{2m}\alpha+\frac{3m\omega^2}{8\alpha}$$

Again, forming the second order equation of α in terms of $E_1(\alpha)$, we get:

$$12\hbar^2\alpha^2-8m\alpha E_1(\alpha)+3m^2\omega^2=0$$

and then,

$$\alpha_{1,2}=\frac{8mE_1(\alpha)\pm\sqrt{(-8mE_1(\alpha))^2-4\times12\hbar^2\times3m^2\omega^2}}{2\times12\hbar^2}$$

$$8mE_1(\alpha)\geq12\hbar m\omega\Rightarrow E_1(\alpha)\geq\frac{3\hbar\omega}{2}$$

which is the same value obtained from the variational method.

Exp. 2

Estimating the ground state energy of the Hydrogen atom using Delta method.

Choosing the trial function in the form of $\psi(r,\theta,\varphi)=e^{-\frac{r}{\alpha}}$, where α is a scale factor, energy is obtained as:

$$E(\alpha)=\frac{\langle\psi|H|\psi\rangle}{\langle\psi|\psi\rangle}=\frac{-\left\langle\psi\left|\frac{\hbar^2}{2m}+\frac{e^2}{r}\right|\psi\right\rangle}{\langle\psi|\psi\rangle}$$

$$\langle\psi|\psi\rangle=\int_0^{+\infty}r^2e^{-\frac{2r}{\alpha}}dr\int_0^{\pi}\sin\theta d\theta\int_0^{2\pi}d\varphi=\pi\alpha^3$$

$$-\left\langle\psi\left|\frac{e^2}{r}\right|\psi\right\rangle=-4\pi e^2\int_0^{+\infty}re^{-\frac{2r}{\alpha}}dr=-\pi e^2\alpha^2$$

$$-\left\langle\psi\left|\frac{\hbar^2}{2m}\nabla^2\right|\psi\right\rangle=\frac{4\pi}{\alpha^2}\frac{\hbar^2}{2m}\int_0^{+\infty}r^2e^{-\frac{2r}{\alpha}}dr=\frac{\hbar^2\pi}{2m}\alpha$$

Now, using Delta method we have:

$$2m\alpha^2E(\alpha)+2m\alpha e^2-\hbar^2=0\rightarrow 8mE(\alpha)\hbar^2\geq-(2me^2)^2$$

$$E(\alpha)\geq-\frac{me^4}{2\hbar^2}$$

that is the same estimated value for the ground state energy of the Hydrogen atom obtained from the variational method.

We should note that if we find $E(\alpha)$ for Hydrogen atom through variational method with the same trial function $\psi(r,\theta,\varphi)=\mathrm{e}^{-\frac{r}{\alpha}}$:

$$E(\alpha)=\frac{\langle\psi|H|\psi\rangle}{\langle\psi|\psi\rangle}\rightarrow E(\alpha)=\frac{\hbar^2}{2m\alpha^2}-\frac{e^2}{\alpha}\rightarrow\frac{\mathrm{d}E}{\mathrm{d}\alpha}=0\rightarrow-\frac{\hbar^2}{m\alpha^3}+\frac{e^2}{\alpha}=0$$

$$\rightarrow\frac{1}{\alpha^2}\left(-\frac{\hbar^2}{m\alpha}+e^2\right)=0\rightarrow\underbrace{E(\alpha)=-\frac{me^4}{2\hbar^2}}_{\alpha=\frac{\hbar^2}{me^2}},\ \underbrace{E(\alpha)=0}_{\alpha=\infty}$$

we find that one of the solutions for α is infinite which is an unphysical solution, since it leads to a zero value for the ground or excited state energy. While, in Delta method, only the physical solution is attained.

Exp. 3

Estimation of the ground state energy of a one-dimensional harmonic oscillator by making use of Delta method with the following two trial functions:

(a) $\psi_0(x,\alpha)=A\mathrm{e}^{-\alpha|x|}$

(b) $\psi_0(x,\alpha)=\frac{A}{x^2+\alpha}$

where α is a positive real number and A is the normalization constant.

(a) We have:

$$E(\alpha)=\frac{\left\langle\psi_0\left|-\frac{\hbar^2}{2m}\frac{\mathrm{d}^2}{\mathrm{d}x^2}+\frac{1}{2}m\omega^2x^2\right|\psi_0\right\rangle}{\langle\psi_0|\psi_0\rangle}$$

The final solution is:

$$2\hbar^2\alpha^4 + m^2\omega^2 = 4m\alpha^2 E(\alpha)$$

$$\alpha^2 = x \rightarrow 2\hbar^2\alpha^4 - 4m\alpha^2 E(\alpha) + m^2\omega^2 = 0$$

$$2\hbar^2 x^2 - 4mxE(\alpha) + m^2\omega^2 = 0$$

$$E(\alpha) \geq \frac{\sqrt{2}}{2}\hbar\omega$$

or

$$E(\alpha) \geq 0.707\hbar\omega\,,$$

which is the ground state energy of a one-dimensional harmonic oscillator with the above defined trial function and is the same result obtained from the variational method.

(b) We have:

$$E(\alpha) = \frac{\langle\psi_0|H|\psi_0\rangle}{\langle\psi_0|\psi_0\rangle} = \frac{\hbar^2}{4m\alpha} + \frac{1}{2}m\omega^2\alpha = \frac{\hbar^2 + 2m^2\omega^2\alpha^2}{4m\alpha}$$

$$\Rightarrow 2m^2\omega^2\alpha^2 + \hbar^2 = 4m\alpha E(\alpha)$$

therefore

$$E(\alpha) \geq \frac{\hbar\omega}{\sqrt{2}}$$

Again, it is quite in accordance with the resulted solution of variational method.

Exp. 4

Applying Delta method, we calculate the ground state energy for a particle of mass m which is bouncing vertically and elastically on a reflecting hard floor where $v(z) = \begin{cases} mgz & z > 0 \\ +\infty & z \leq 0 \end{cases}$ and g is the gravitational constant.

Choosing the trial function $\psi_0(z,x) = Aze^{-\alpha z}$, where α is a parameter and A is the normalization constant, we can show $A = z\alpha^{\frac{3}{2}}$ and therefore:

$$\psi_0(z,\alpha) = 2\sqrt{\alpha^3}\, ze^{-\alpha z}$$

$$E(\alpha) = \langle \psi_0 | H | \psi_0 \rangle = 4\alpha^3 \int_0^{+\infty} ze^{-\alpha z}\left[-\frac{\hbar^2}{2m}\frac{d^2}{dz^2} + mgz\right] ze^{-\alpha z}dz$$

$$\hbar^2\alpha^3 + 3m^2g - 2m\alpha E(\alpha) = 0$$

In the above third order equation of α in terms of $E(\alpha)$, using Equations (2) to (8) we have:

$$\begin{cases} p = \frac{1}{3}\left[\frac{3(-2mE)}{\hbar^2}\right] - \frac{0}{\hbar^2} = -\frac{2mE}{\hbar^2} \\ q = \frac{1}{27}\left(\frac{2(0)^3}{a^3} - \frac{2(0)c}{a^2} + \frac{(27)(3m^2g)}{\hbar^2}\right) = \frac{1}{27}\left((27)\frac{3m^2g}{\hbar^2}\right) = \frac{3m^2g}{\hbar^2} \\ D = \left(\frac{-2mE}{3\hbar^2}\right)^3 + \left(\frac{3m^2g}{2\hbar^2}\right)^2 \end{cases}$$

The third order equation has some defined solutions due to $D \le 0$ or $D > 0$ or $D = 0$. The physical solution is obtained for $D \le 0$ as:

$$-\left(\frac{2mE}{3\hbar^2}\right)^3 + \left(\frac{3m^2g}{2\hbar^2}\right)^2 \le 0$$

$$\frac{2mE}{3\hbar^2} \ge \left(\frac{3}{2}\right)^{\frac{2}{3}} \frac{m^{\frac{4}{3}}(g)^{\frac{2}{3}}}{\hbar^{\frac{4}{3}}}$$

$$E \ge \left(\frac{3}{2}\right)\left(\frac{9}{4}\right)^{\frac{1}{3}} \left(mg^2\hbar^2\right)^{\frac{1}{3}}$$

which is consistent with the solution obtained from variational method, i.e.

$$E \ge \left(\frac{3}{2}\right)\left(\frac{9}{2}\right)^{\frac{1}{3}} \left(\frac{1}{2}mg^2\hbar^2\right)^{\frac{1}{3}}.$$

Exp. 5

Estimation of the ground state energy and the corresponding wave function of a system consisting of two identical particles of spin $\frac{1}{2}$ that are confined to an isotropic three-dimensional harmonic oscillator potential of frequency ω, using delta method, when the two particles do not interact.

Choosing the trial function in the form of $\psi(\vec{r}_1,\vec{r}_2,\vec{s}_1,\vec{s}_e) = A e^{-\alpha(r_1+r_2)}\psi$, we have:

$$H = -\frac{\hbar^2}{2m}\nabla_1^2 - \frac{\hbar^2}{2m}\nabla_2^2 + \frac{1}{2}m\omega^2 r_1^2 + \frac{1}{2}m\omega^2 r_2^2$$

$$1 = \langle\psi|\psi\rangle = |A|^2\left[4\pi\int_0^{+\infty} e^{-2\alpha r_1} r_1^2 dr_1\right]\left[4\pi\int_0^{+\infty} e^{-2\alpha r_2} r_2^2 dr_2\right] = |A|^2 \frac{4\pi}{4\alpha^3}\frac{4\pi}{4\alpha^3} = \frac{|A|^2 \pi^2}{\alpha^6}$$

$$|A|^2 = \left(\frac{\alpha^3}{\pi}\right)^2 = \frac{\alpha^6}{\pi^2}$$

$$\langle\psi|H|\psi\rangle = 2|A|^2 \int_v e^{-2\alpha r_2} dv_2 \int_v e^{-\alpha r_1}\left[-\frac{\hbar^2}{2m}\nabla_1^2 + \frac{1}{2}m\omega^2 r_1^2\right] e^{-\alpha r_1} dv_1$$

$$= \frac{12m\omega^2}{\alpha^2} + \frac{4\hbar^2\alpha^2}{m} = E$$

Applying Delta method, we get:

$$4\hbar^2\alpha^4 - Em\alpha^2 + 12m^2\omega^2 = 0,$$

which with

$$\alpha^2 = x \rightarrow 4\hbar^2 x^2 - Emx^2 + 12m^2\omega^2 = 0$$

we will encounter the following condition:

$$E \geq 8\sqrt{3}\hbar\omega,$$

that is the same energy of the ground state obtained from the variational method.

5. CONCLUSION

In this paper, we used an alternative method of finding the ground and excited energies state of stationary states. Basically, it is the same as the variational method, where a time-independent Hamiltonian $\hat{H}$ with no exact solution is exploited, however, in this new method, we used a trial function and calculated the value of energy E_n in terms of a defined parameter α. Therefore, the estimation of the upper bound values and the excited state energies, is based on applying the Delta conditions $\Delta \geq 0$ and $D \leq 0$, which are included respectively, in appropriate second and third order equations. Presenting some examples, we showed that the corresponding results which have been obtained within this new method are in consistence with those derived from the traditional variational method. It is remarkable that all the problems to be solved through variational method, could be solved through Delta method and lead to exactly the same result with the difference that Delta method is simpler and the calculation procedures are less than that of the variational method.

REFERENCES

1. Zettili, N. (2009) Quantum Mechanics and Applications. John Wiley & Sons, Hoboken.
2. Domínguez-Hernández, S. and Fernández C, D.J. (2011) Rosen-Morse Potential and Its Supersymmetric Partners. International Journal of Theoretical Physics, 50, 1993-2001.
3. Díaz, J.I., Negro, J., Nieto, L.M. and Rosas-Ortiz, O. (1999) The Supersymmetric Modified Pöschl-Teller and Delta-Well Potentials. Journal of Physics A: Mathematical and General, 32, 8447-8460.
4. Negro, J., Nieto, L.M. and Rosas-Ortiz, O. (2000) Confluent Hypergeometric Equations and Related Solvable Potentials in Quantum Mechanics. Journal of Mathematical Physics, 41, 7964-7996.
5. Contreras-Astorga, A. and Fernández C, D.J. (2008) Supersymmetric Partners of the Trigonometric Pöschl-Teller Potentials. Journal of Physics A: Mathematical and General, 41, Article ID: 475303.

6. Compean, C.B. and Kirchbach, M. (2006) The Trigonometric Rosen-Morse Potential in Supersymmetric Quantum Mechanics and Its Exact Solutions. Journal of Physics A: Mathematical and General, 39, 547-557.

7. Karayer, H., Demirhan, D. and Büyükkilic, F. (2015) Extension of Nikivorov-Uvarov Method for the Solution of Heun Equation. Journal of Mathematical Physics, 56, 1-22.

8. Yasuk, F., Berkdemir, C. and Berkdemir, A. (2005) Exact Solutions of the Schrodinger Equation with Non-Central Potential by Nikiforov-Uvarov Method. Journal of Physics A: Mathematical and General, 38, Article No. 29.

9. Gönül, B. and Köksal, K. (2007) A Search on the Nikiforov-Uvarov Formalism. Physica Scripta, 75, 686-690.

10. Quesne, C. (2013) Extending Romanowski Polynomials in Quantum Mechanics. Journal of Mathematical Physics, 54, Article ID: 122103.

11. Weber, H.J. (2007) Connection between Real Polynomial Solutions of Hypergeometric-Type Differential Equations with Rodrigues Formula. Central European Journal of Mathematics, 5, 415-427.

12. Weber, H.J. (2007) Connection between Romanowski and Other Polynomials. Central European Journal of Mathematics, 5, 581-595.

13. Quesne, C. (2008) Exceptional Orthogonal Polynomials, Exactly Solvable Potentials and Supersymmetry. Journal of Physics A: Mathematical and Theoretical, 41, Article ID: 392001.

14. Quesne, C. (2009) Solvable Rational Potentials and Exceptional Orthogonal Polynomials in Supersymmetric Quantum Mechanics. SIGMA, 5, 1-24.

15. Quesne, C. (2011) Higher Order SUSY, Exactly Solvable Potentials, and Exceptional Orthogonal Polynomials. Modern Physics Letters A, 26, 1843-1852.

16. Quesne, C. (2011) Rationally-Extended Radial Oscillators and Laguerre Exceptional Orthogonal Polynomials in kth-Order SUSYQM. International Journal of Modern Physics A, 26, 5337-5347.

17. Quesne, C. (2012) Revisiting (Quasi-)Exactly Solvable Rational Extensions of the Morse Potential. International Journal of Modern Physics A, 27, Article ID: 1250073.

18. Quesne, C. (2012) Novel Enlarged Shape Invariance Property and Exactly Solvable Rational Extensions of the Rosen-Morse II and Eckart Potentials. SIGMA, 8, 1-19.

19. Odake, S. and Sasaki, R. (2009) Infinitely Many Shape Invariant Potentials and the New Orthogonal Polynomials. Physics Letters B, 679, 414-417.

20. Odake, S. and Sasaki, R. (2013) Extensions of Solvable Potentials with Finitely Many Discrete Eigenstates. Journal of Physics A: Mathematical and Theoretical, 46, Article ID: 235205.

21. Odake, S. and Sasaki, R. (2011) Exactly Solvable Quantum Mechanics and Infinite Families of Multi-Indexed Orthogonal Polynomials. Physics Letters B, 702, 164-170.

22. Grandati, Y. (2011) Solvable Rational Extensions of the Morse and Kepler-Coulomb Potentials. Journal of Mathematical Physics, 52, Article ID: 103505.

23. Ho, C.-L. (2011) Prepotential Approach to Solvable Rational Potentials and Exceptional Orthogonal Polynomials. Progress of Theoretical Physics, 126, 185-201.

24. Woan, G. (2000) The Cambridge Handbook of Physics Formulas. Cambridge University Press, Cambridge, UK.

7
CHAPTER

QUANTUM-SQUEEZING EFFECTS OF STRAINED MULTILAYER GRAPHENE NEMS

Yang Xu[1*], Sheping Yan[1], Zhonghe Jin[1*] and Yuelin Wang[2]

[1]Department of Information Science and Electronic Engineering, Zhejiang University, Hangzhou 310027, China

[2]State Key Laboratory of Transducer Technology, Shanghai Institute of Metallurgy Chinese Academy of Sciences, Shanghai 100050, China

ABSTRACT

Quantum squeezing can improve the ultimate measurement precision by squeezing one desired fluctuation of the two physical quantities in Heisenberg relation. We propose a scheme to obtain squeezed states through graphene nanoelectromechanical system (NEMS) taking advantage of their thin thickness in principle. Two key criteria of achieving squeezing states, zero-point displacement uncertainty and squeezing factor of strained multilayer graphene NEMS, are studied. Our research promotes the measured precision limit of graphene-based nano-transducers by reducing quantum noises through squeezed states.

INTRODUCTION

The Heisenberg uncertainty principle, or the standard quantum limit [1, 2], imposes an intrinsic limitation on the ultimate sensitivity of quantum measurement systems, such as atomic forces [3], infinitesimal displacement [4], and gravitational-wave [5] detections. When detecting very weak physical quantities, the mechanical motion of a nano-resonator or nanoelectromechanical system (NEMS) is comparable to the intrinsic fluctuations of the systems, including thermal and quantum fluctuations. Thermal fluctuation can be reduced by decreasing the temperature to a few mK, while quantum fluctuation, the quantum limit determined by Heisenberg relation, is not directly dependent on the temperature. Quantum squeezing is an efficient way to decrease the system quantum [6–8]. Thermomechanical noise squeezing has been studied by Rugar and Grutter [9], where the resonator motion in the fundamental mode was parametrically squeezed in one quadrature by periodically modulating the effective spring constant at twice its resonance frequency. Subsequently, Suh et al. [10] have successfully achieved parametric amplification and back-action noise squeezing using a qubit-coupled nanoresonator.

To study quantum-squeezing effects in mechanical systems, zero-point displacement uncertainty, Δx_{zp}, the best achievable measurement precision, is introduced. In classical mechanics, the complex amplitudes, $X = X_1 + iX_2$, where X_1 and X_2 are the real and imaginary parts of complex amplitudes respectively, can be obtained with complete precision. In quantum mechanics, X_1 and X_2 do not commute, with the commutator $[X_1, X_2] = i\hbar/M_{eff}\, w$, and satisfy the uncertainty relationship $\Delta X_1 \Delta X_2 \geq (\hbar/2M_{eff}\, w)^{1/2}$. Here, $\hbar$ is the Planck constant divided by 2π, $M_{eff} = 0.375\rho LWh/2$ is the effective motional double-clamped film mass [11, 12], ρ is the volumetric mass density, L, W, and h are the length, width, and thickness of the film, respectively, and $w = 2f_0$ is the fundamental flexural mode angular frequency with

$$f_0 = \{[A(E/\rho)^{1/2} h/L^2]^2 + A^2 0.57 T_s/\rho L^2 Wh\}^{1/2}, \quad (1)$$

where E is the Young's modulus of the material, T_s is the tension on the film, A is 0.162 for a cantilever and A is 1.03 for a double-clamped film [13]. Therefore, Δx_{zp} of the fundamental mode of a NEMS device with a double-clamped film can be given by $\Delta x_{zp} = \Delta X_1 = \Delta X_2 = (\hbar/2M_{eff}\, w)^{1/2}$. In a mechanical system, quantum squeezing can reduce the displacement uncertainty Δx_{zp}.

Recently, free-standing graphene membranes have been fabricated [14], providing an excellent platform to study quantum-squeezing effects in mechanical systems. Meanwhile, a graphene membrane is sensitive to external influences, such as atomic forces or infinitesimal mass (e.g., 10^{-21} g) due to its atomic thickness. Although graphene films can be used to detect very infinitesimal physical quantities, the quantum fluctuation noise Δx_{zp} of graphene NEMS devices (approx. 10^{-2} nm), could easily surpass the magnitudes of signals caused by external influences. Thus, quantum squeezing becomes necessary to improve the ultimate precision of graphene-based transducers with ultra-high sensitivity. In this study, we have studied quantum-squeezing effects of strained multilayer graphene NEMS based on experimental devices proposed by Chen et al. [15].

RESULTS

Displacement uncertainty of graphene NEMS

A typical NEMS device with a double-clamped free-standing graphene membrane is schematically shown in Figure 1. The substrate is doped Si with high conductivity, and the middle layer is SiO_2 insulator. A pump voltage can be applied between the membrane and the substrate. The experimental data of the devices are used in our simulation [15]. For graphene, we use a Young's modulus of $E = 1.03 \times 10^{12}$ Pa, volumetric mass density of $\rho = 2200$ kg/m^3, based on previous theories and scanning tunneling microscope experiments [13, 15, 16].

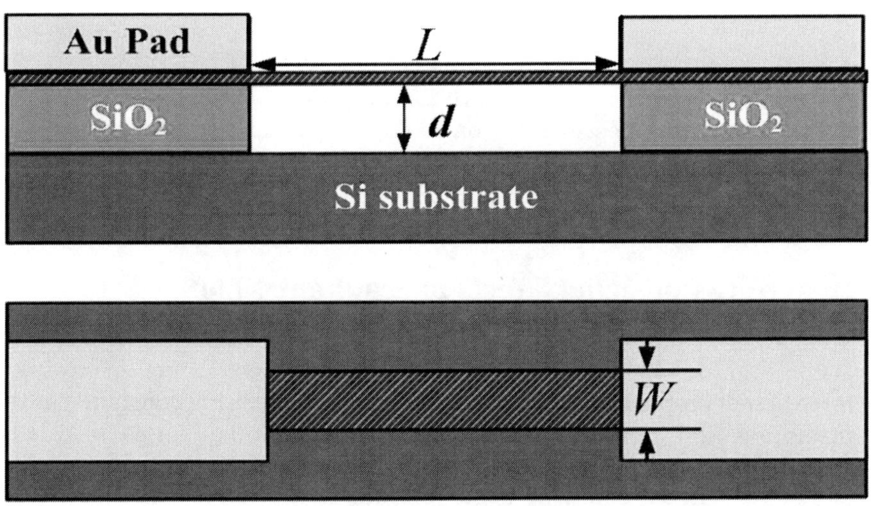

Figure 1. Schematic of a double-clamped graphene nems device.

In graphene sensors and transducers, to detect the molecular adsorbates or electrostatic forces, a strain ε will be generated in the graphene film [15, 17]. When a strain exists in a graphene film, the tension T_s in Equation 1 can be deduced as $T_s = ES\varepsilon = EWh\varepsilon$. The zero-point displacement uncertainty of the strained graphene film is given by

$$\Delta x_{zp} = \sqrt{\hbar/2M_{eff}w} = \sqrt{\hbar/\{2.94\pi\rho' LWh[1.03^2 h^2 E/(L^4\rho') + 0.6047E\varepsilon/(\rho' L^2)]^{1/2}\}} \quad (2)$$

where ρ' represents the effective volumetric mass density of graphene film after applying strain. The typical measured strains in [15] are $\varepsilon = 4 \times 10^{-5}$ when $\rho' = 4\rho$ and $\varepsilon = 2 \times 10^{-4}$ when $\rho' = 6\rho$. Based on Equation 2, measurable Δx_{zp} of the strained multilayer graphene films of various sizes are shown in Figure 2, and typical Δx_{zp} values of graphene NEMS under various ε are summarized in Table 1.

Table 1. Calculated Δx_{zp} (10^{-4}nm) of monolayer (Mon), bilayer (Bi), and trilayer (Tri) graphene versus strain ε (L = 1.1 μm, W = 0.2 μm)

$\varepsilon = 0$			$\varepsilon = 4 \times 10^{-5}$			$\varepsilon = 2 \times 10^{-4}$		
Mon	Bi	Tri	Mon	Bi	Tri	Mon	Bi	Tri
34.0	17.0	11.3	6.05	4.23	3.39	3.67	2.59	2.10

According to the results in Figure 2 and Table 1, we find $\Delta x_{zp}^{\text{large strain}} < \Delta x_{zp}^{\text{small strain}}$; one possible reason is that larger applied strain results in smaller fundamental angular frequency and Δx_{zp}, therefore, the quantum noise can be reduced.

Quantum-squeezing effects of graphene NEMS

To analyze quantum-squeezing effects in graphene NEMS devices, a back-action-evading circuit model is used to suppress the direct electrostatic force acting on the film and modulate the effective spring constant k of the membrane film. Two assumptions are used, namely, the film width W is on the micrometer scale and $X_1 >> d$, where d is the distance between the film and the substrate. Applying a pump voltage $V_m(t) = V[1+ \sin(2w_m t + \theta)]$, between the membrane film and the substrate, the spring constant k will have a sinusoidal modulation $k_m(t)$, which is given by $k_m(t) = \sin(2w_m t + \theta)C_T V^2/2d^2$, where C_T is the total capacitance composed of

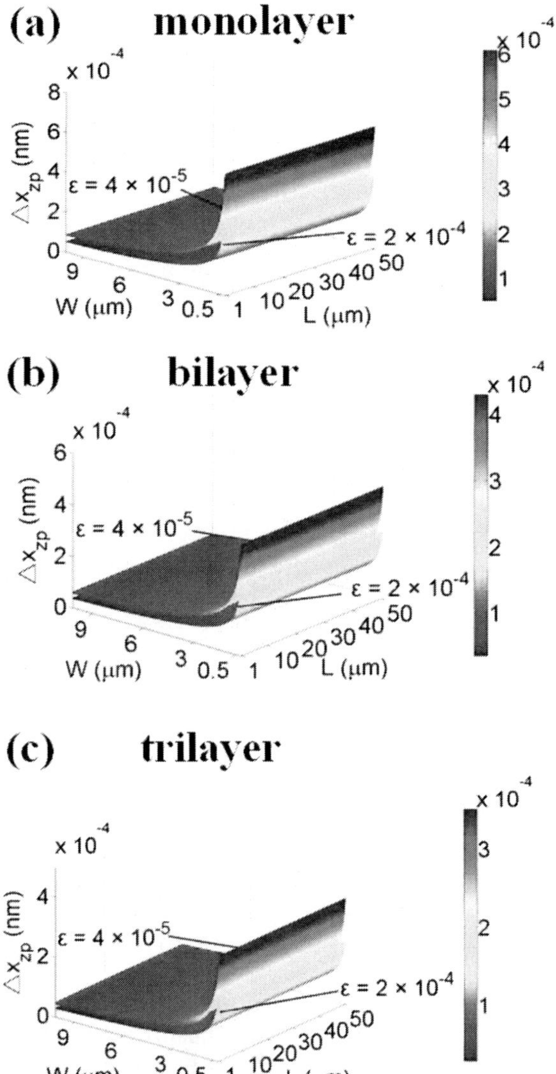

Figure 2. **Δx_{ZP} VERSUS MULTILAYER GRAPHENE FILM SIZES WITH STRAINS**. **(A)** Monolayer graphene. **(B)** Bilayer graphene. **(C)** Trilayer graphene.

structure capacitance C_0, quantum capacitance C_q, and screen capacitance C_s in series [18]. The quantum capacitance C_q and screen capacitance C_s cannot be neglected [18–20] owing to a graphene film thickness on the atomic scale. The quantum capacitance of monolayer graphene [21, 22] is $C_q^{\text{monolayer}} = 2e^2 n^{1/2}/(\hbar v_F \pi^{1/2})$, where n is the carrier concentration, e is the elementary charge, and $v_F \approx c/300$, where c is the velocity of light, with bilayer $C_q^{\text{bilayer}} = 2 \times 0.037 m_e e^2/\pi\hbar^2$, and trilayer $C_q^{\text{trilayer}} = 2 \times 0.052 m_e e^2/\pi\hbar^2$, where m_e is the electron mass [23].

Pumping the graphene membrane film from an initial thermal equilibrium state at frequency $w_m = w$, the variance of the complex amplitudes, $\Delta X^2_{1,2}(t, \theta)$, are given by [24]

$$\Delta X^2_{1,2}(t,\theta) = (\hbar/2M_{\text{eff}} w)(2N+1)\exp(-t/\tau)[ch(2\eta t) \mp \cos\theta sh(2\eta t) + \tau^{-1}(I_c \pm \cos\theta I_d)], \tag{3}$$

where $I_c = \int_0^t e^{t/\tau} ch[2\eta(\delta - t)]d\delta$, $I_d = \int_0^t e^{t/\tau} sh[2\eta(\delta - t)]d\delta$, $N = [\exp(\hbar w/k_B T) - 1]^{-1}$ is the average number of quanta at absolute temperature T and frequency w, k_B is the Boltzmann constant, $\tau = Q/w$ is the relaxation time of the mechanical vibration, Q is the quality factor of the NEMS, and $\eta = C_T V^2/8d^2 M_{\text{eff}} w_m$. When $\theta = 0$, a maximum modulation state, namely, the best quantum-squeezed state, can be reached [9, 21], and ΔX_1 can be simplified as $\Delta X_1(t) = [(\hbar/2M_{\text{eff}} w_a)(2N + 1)(\tau^{-1} + 2\eta)^{-1}(\tau^{-1} + 2\eta \exp(-\tau^{-1} + 2\eta)t)]^{1/2}$. As $t \to \infty$, the maximum squeezing of ΔX_1 is always finite, with expression of $\Delta X_1(t \to \infty) \approx [\hbar(2N + 1)(1 + 2Q\eta)^{-1}/2M_{\text{eff}} w]^{1/2}$. The squeezing factor R, defined as $R = \Delta X_1/\Delta x_{zp} = \Delta X_1/(\hbar/2M_{\text{eff}} w)^{1/2}$, can be expressed as

$$R = \sqrt{\frac{2/\{\exp[\hbar(k_B T)^{-1} 2\pi(1.03^2 h^2 E/(L^4\rho') + 0.6047E\varepsilon/(\rho' L^2))^{1/2}] - 1\} + 1}{1 + QC_T V^2(4d^2)^{-1}\{[2\pi\rho' LWh(1.03^2 h^2 E/(L^4\rho') + 0.6047E\varepsilon/(\rho' L^2))]^{1/2}\}^{-1}}}. \tag{4}$$

where ε is the strain applied on the graphene film. In order to achieve quantum squeezing, R must be less than 1. According to Equation 4, R values of monolayer and bilayer graphene films with various dimensions, strain ε, and applied voltages at $T = 300$ K and $T = 5$ K have been shown in Figure 3. Quantum squeezing is achievable in the region $\log R < 0$ at $T = 5$ K. As shown in Figure 3, the applied strain increases the R values because of the increased fundamental angular frequency and the decreased Δx_{zp} caused by strain, which makes squeezing conditions more difficult to reach. Figure 4a has shown that ΔX_1 changes with applied voltages at $T = 5$ K, the red line represents the uncertainties of X_1 and the dashed reference line is $\Delta X = \Delta x_{zp}$. As shown in Figure 4a, applying a voltage larger than 100 mV, we can obtain $\Delta X_1 < \Delta x_{zp}$, which means that the displacement uncertainty is squeezed, and the quantum squeezing is achieved. Some typical *R* values of monolayer graphene film, obtained by varying the applied voltage *V*, such as strain ε, have been listed in Table 2 (with *T* = 300 K and *Q* = 125) and Table 3 (with *T* = 5 K and *Q* = 14000). As shown in Tables 2 and 3 and Figure 3, lowering the temperature to 5 K can dramatically decrease the *R* values. The lower the temperature, the larger the quality factor *Q*, which makes the squeezing effects stronger.

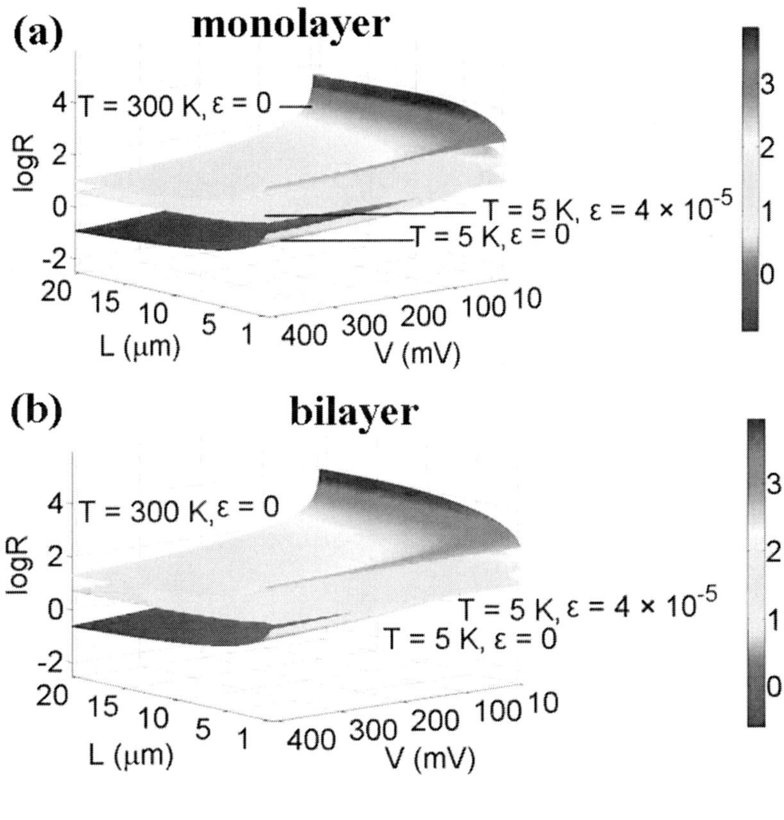

Figure 3. Log *r* versus applied voltages for graphene film structures at ***t*** = 300 k with ***q*** = 125 and ***t*** = 5 k with ***q*** = 14000. **(A)**Monolayer graphene and **(B)** bilayer graphene.

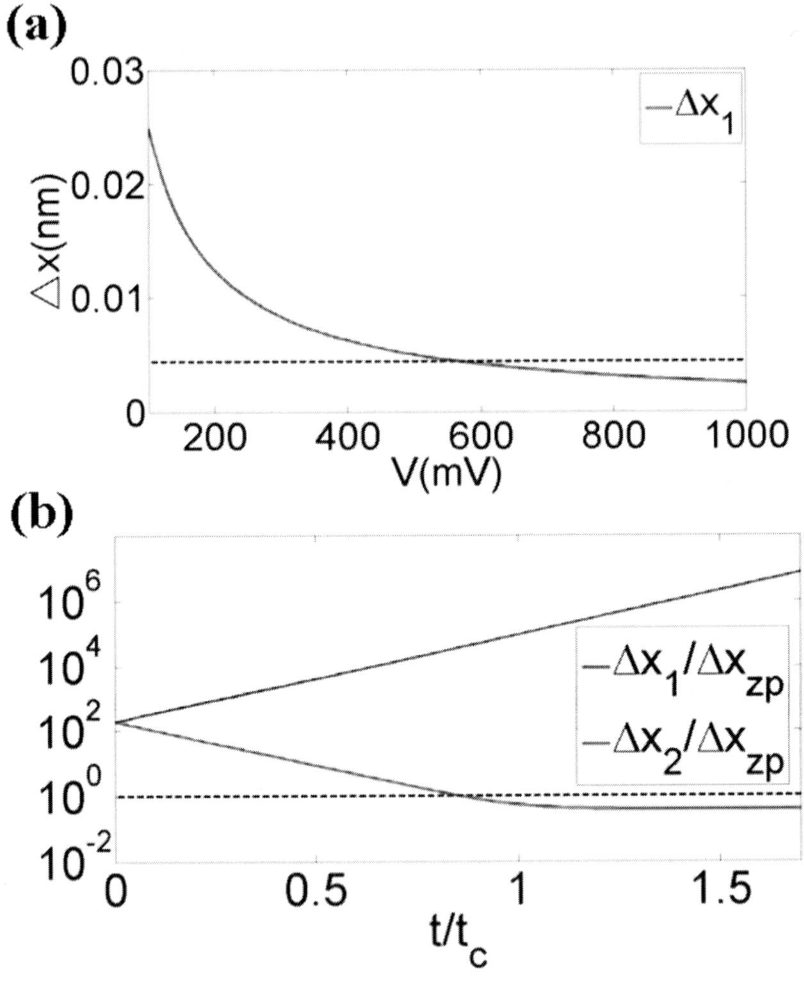

Figure 4. (A) $\Delta \boldsymbol{X}_1$ VERSUS APPLIED VOLTAGES OF GRAPHENE FILM AND THE DASHED REFERENCE LINE IS $\Delta \boldsymbol{X} = \Delta \boldsymbol{x}_{ZP}$. (B) Time dependences of $\Delta X_1$and ΔX_2, which are expressed in units of Δx_{zp}, where time is in units of t_{ct}, $\theta = 0$, and the dashed reference line is $\Delta X = \Delta x_{zp}$. $L = 1.1$ μm, $W = 0.2$ μm, $d = 0.1$ μm, $T = 5$ K, $Q = 14000$, and $V = 2.5$V.

Table 2. R values of monolayer graphene versus various strain ε and voltage V (L = 1.1 μm, W = 0.2 μm, and T = 300 K with Q = 125)

	$\varepsilon = 0$	$\varepsilon = 4 \times 10^{-5}$	$\varepsilon = 2 \times 10^{-4}$
V = 2 V	38.33	198.15	259.14
V = 10 V	7.669	42.84	69.86

Table 3. R values of monolayer graphene versus various strain ε and voltage V (L = 1.1 μm, W − 0.2 μm, and T = 5 K with Q = 14000)

	$\varepsilon = 0$	$\varepsilon = 4 \times 10^{-5}$	$\varepsilon = 2 \times 10^{-4}$
V = 2 V	0.468	2.620	4.319
V = 10 V	0.0936	0.524	0.867

In contrast to the previous squeezing analysis proposed by Rugar and Grutter [9], in which steady-state solutions have been assumed and the minimum R is 1/2, we use time-dependent pumping techniques to prevent X_2 from growing without bound as $t \rightarrow \infty$, which should be terminated after the characteristic time $t_{ct} = \ln(QC_T V^2/4M_{eff} w^2 d^2)4M_{eff} wd^2/C_T V^2$, when R achieves its limiting value. Therefore, we have no upper bound on R. Figure 4b has shown the time dependence of ΔX_1 and ΔX_2 in units of t_{ct}, and the quantum squeezing of the monolayer graphene NEMS has reached the limiting value after one t_{ct} time. Also, to make the required heat of conversion from mechanical energy negligible during the pump stage, $t_{ct} << \tau$ must be satisfied. We find $t_{ct}/\tau \approx 1.45 \times 10^{-5}$ for the monolayer graphene parameters considered in the text.

DISCUSSION

The ordering relation of Δx_{zp} for multilayer graphene is $\Delta x_{zp}^{trilayer} < \Delta x_{zp}^{bilayer} < \Delta x_{zp}^{monolayer}$ shown in Figure 5a, as the zero-point displacement uncertainty is inversely proportional to the film thickness. Squeezing factors R of multilayer graphene films follow the ordering relation; $R_{trilayer} > R_{bilayer} > R_{monolayer}$, as shown in Figure 5b, as R is proportional to the thickness of the graphene film. The thicker the film, the more difficult it is to achieve a quantum-squeezed state, which also explains why traditional NEMS could not achieve quantum squeezing due to their thickness of several hundred nanometers.

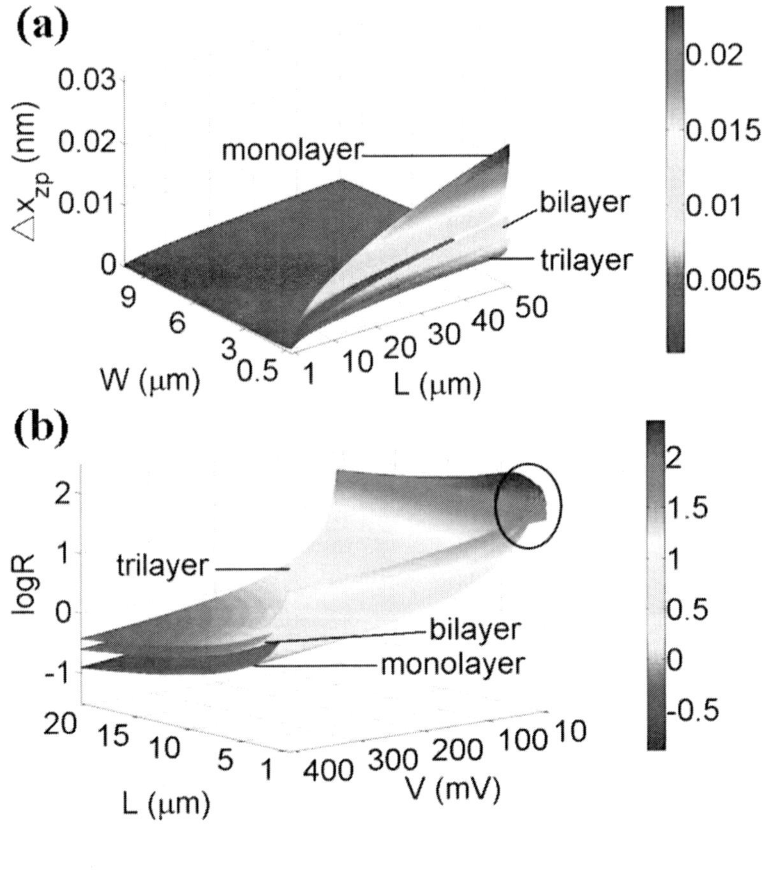

Figure 5. (A) Δx_{ZP} VERSUS VARIOUS GRAPHENE FILM SIZES. (B) Log *R* versus multilayer graphene film lengths and applied voltages at $T = 5$ K

For a clear view of squeezing factor *R* as a function of film length *L*, 2D curves from Figure 5b are presented in Figure 6. It is found that *R* approaches unity as *L* approaches zero, while *R* tends to be zero as *L* approaches infinity as shown in Figure 6a,b. It explains why *R* has some kinked regions, shown in the upper right part of Figure 5b with black circle, when the graphene film length is on the nanometer scale shown in Figure 3.

To realize quantum squeezing, the graphene film length should be in the order of a few micrometers and the applied voltage V should not be as small as several mV, shown in Figure 6b. As $L \rightarrow 0$, where the graphene film can be modeled as a quantum dot, the voltage must be as large as a few volts to modulate the film to achieve quantum squeezing. As $L \rightarrow \infty$, where graphene films can be modeled as a 1D chain, the displacement uncertainty would be on the nanometer scale so that even a few mV of pumping voltage can modulate the film to achieve quantum squeezing easily.

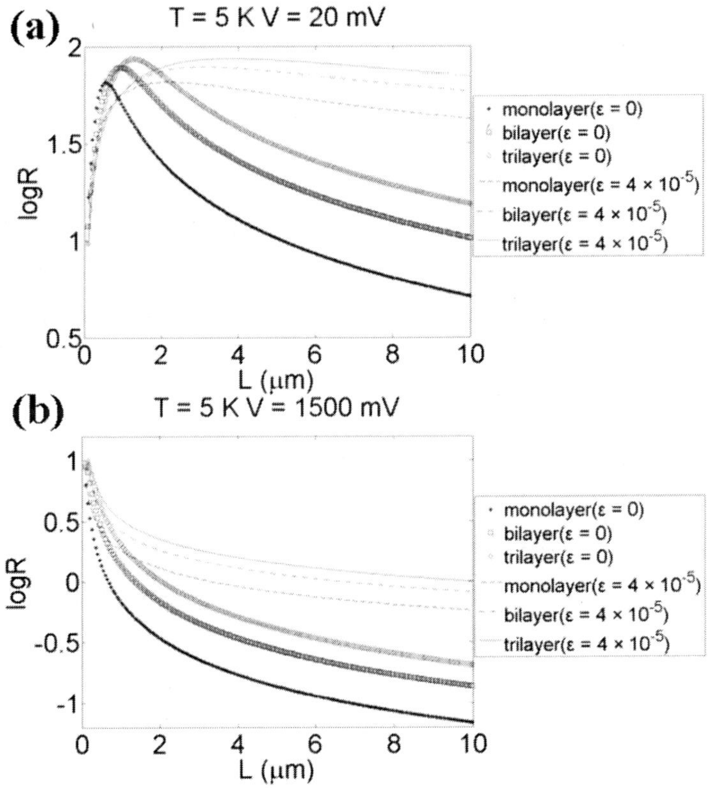

Figure 6. ***R*** **VERSUS** ***L*** **WITH** $\varepsilon = 0.4 \times 10^{-5}$**, AND** ***V*** **= 20 MV, 1.5 V.**

By choosing the dimensions of a typical monolayer graphene NEMS device in [15] with L = 1.1 μm, W = 0.2 μm, T = 5 K, Q = 14000, V = 2.5 V, and $\varepsilon = 0$, we obtain Δx_{zp} = 0.0034 nm and R = 0.374. After considering quantum squeezing effects based on our simulation, Δx_{zp} can be reduced to 0.0013 nm. With a length of 20 μm, Δx_{zp} can be as large as 0.0145 nm, a radio-frequency single-electron-transistor detection system can in principle attain such sensitivities [25]. In order to verify the quantum squeezing effects, a displacement detection scheme need be developed.

CONCLUSIONS

In conclusion, we presented systematic studies of zero-point displacement uncertainty and quantum squeezing effects in strained multilayer graphene NEMS as a function of the film dimensions *L*, *W*, *h*, temperature *T*, applied voltage *V*, and strain ε applied on the film. We found that zero-point displacement uncertainty Δx_{zp} of strained graphene NEMS is inversely proportional to the thickness of graphene and the strain applied on graphene. By considering quantum capacitance, a series of squeezing factor *R* values have been obtained based on the model, with $R_{monolayer} < R_{bilayer} < R_{trilayer}$ and $R_{small\ strain} < R_{large\ strain}$ being found. Furthermore, high-sensitivity graphene-based nano-transducers can be developed based on quantum squeezing.

DECLARATIONS

Acknowledgements

The authors gratefully acknowledge Prof. Raphael Tsu at UNCC, Prof. Jean-Pierre Leburton at UIUC, Prof. Yuanbo Zhang at Fudan University, Prof. Jack Luo at University of Bolton, and Prof. Bin Yu at SUNY for fruitful discussions and comments. This study is supported by the National Science Foundation of China (Grant No. 61006077) and the National Basic Research Program of China (Grant Nos. 2007CB613405 and 2011CB309501). Dr. Y. Xu is also supported by the Excellent Young Faculty Awards Program (Zijin Plan) at Zhejiang University and the Specialized Research Fund for the Doctoral Program of Higher Education (SRFDP with Grant No. 20100101120045).

Competing interests

The authors declare that they have no competing interests.

Authors' contributions

Both SY and YX designed and conducted all the works and drafted the manuscript. Both ZJ and YW have read and approved the final manuscript.

REFERENCES

1. LaHaye MD, Buu O, Camarota B, Schwab KC: Approaching the quantum limit of a nanomechanical resonator. Science 2004, 304: 74–77. 10.1126/science.1094419

2. Blencowe M: Nanomechanical quantum limits. Science 2004, 304: 56–57. 10.1126/science.1095768

3. Caves CM, Thorne KS, Drever RWP, Sandberg VD, Zimmermann M: ON the measurement of a weak classical force coulped to a quantum-mechanical oscillator. I. Issues of principle. Rev Mod Phys 1980, 52: 341–392. 10.1103/RevModPhys.52.341

4. Mozyrsky D, Martin I, Hastings MB: Quantum-limited sensitivity of single-electron-transistor-based displacement detectors. Phys Rev Lett2004, 92: 083103.

5. Hollenhorst JN: Quantum limits on resonant-mass gravitational-radiation detectors. Phys Rev D 1979, 19: 1669–1679. 10.1103/PhysRevD.19.1669

6. Blencowe M: Quantum electromechanical systems. Phys Rep Rev Sec Phys Lett 2004, 395: 159–222.

7. Giovannetti V, Lloyd S, Maccone L: Quantum-enhanced measurements: beating the standard quantum limit. Science 2004, 306: 1330–1336. 10.1126/science.1104149

8. Blencowe MP, Wybourne MN: Quantum squeezing of mechanical motion for micron-sized cantilevers. Physica B 2000, 280: 555–556. 10.1016/S0921-4526(99)01862-1

9. Rugar D, Grutter P: Mechanical parametric amplification and thermomechanical noise squeezing. Phys Rev Lett 1991, 67: 699–702. 10.1103/PhysRevLett.67.699

10. Suh J, LaHaye MD, Echternach PM, Schwab KC, Roukes ML: Parametric amplification and back-action noise squeezing by a

qubit-coupled nanoresonator. Nano Lett 2010, 10: 3990–3994. 10.1021/nl101844r

11. Ekinci KL, Yang YT, Roukes ML: Ultimate limits to inertial mass sensing based upon nanoelectromechanical systems. J Appl Phys 2004, 95: 2682–2689. 10.1063/1.1642738

12. Ekinci KL, Roukes ML: Nanoelectromechanical systems. Rev Sci Instrum 2005, 76: 061101. 10.1063/1.1927327

13. Bunch JS, van der Zande AM, Verbridge SS, Frank IW, Tanenbaum DM, Parpia JM, Craighead HG, McEuen PL: Electromechanical resonators from graphene sheets. Science 2007, 315: 490–493. 10.1126/science.1136836

14. Novoselov KS, Geim AK, Morozov SV, Jiang D, Zhang Y, Dubonos SV, Grigorieva IV, Firsov AA: Electric field effect in atomically thin carbon films. Science 2004, 306: 666–669. 10.1126/science.1102896

15. Chen CY, Rosenblatt S, Bolotin KI, Kalb W, Kim P, Kymissis I, Stormer HL, Heinz TF, Hone J: Performance of monolayer graphene nanomechanical resonators with electrical readout. Nat Nanotechnol 2009, 4: 861–867. 10.1038/nnano.2009.267

16. Ni ZH, Wang HM, Kasim J, Fan HM, Yu T, Wu YH, Feng YP, Shen ZX: Graphene thickness determination using reflection and contrast spectroscopy. Nano Lett 2007, 7: 2758–2763. 10.1021/nl071254m

17. Lee C, Wei XD, Kysar JW, Hone J: Measurement of the elastic properties and intrinsic strength of monolayer graphene. Science 2008, 321: 385–388. 10.1126/science.1157996

18. Xu Y, Aluru NR: Pull-in/out analysis of nano/microelectromechanical switches with defective oxide layers. Appl Phys Lett 2009, 95: 073112. 10.1063/1.3211111

19. Tang Z, Xu Y, Li G, Aluru NR: Physical models for coupled electromechanical analysis of silicon nanoelectromechanical systems. J Appl Phys 2005, 97: 114304. 10.1063/1.1897483

20. Xu Y, Aluru NR: Multiscale electrostatic analysis of silicon nanoelectromechanical systems (NEMS) via heterogeneous quantum models.Phys Rev B 2008, 77: 075313.

21. Fang T, Konar A, Xing HL, Jena D: Carrier statistics and quantum capacitance of graphene sheets and ribbons. Appl Phys Lett 2007, 91: 092109. 10.1063/1.2776887

22. Xia JL, Chen F, Li JH, Tao NJ: Measurement of the quantum capacitance of graphene. Nat Nanotechnol 2009, 4: 505–509. 10.1038/nnano.2009.177

23. Koshino M, Ando T: Orbital diamagnetism in multilayer graphenes: systematic study with the effective mass approximation. Phys Rev B2007, 76: 085425.

24. Grishchuk LP, Sazhin MV: Squeezed quantum states of a harmonic-oscillator in the problem of detecting gravitational-waves. Zh Eksp Teor Fiz 1983, 84: 1937–1950.

25. Turin VO, Korotkov AN: Analysis of the radio-frequency single-electron transistor with large quality factor. Appl Phys Lett 2003, 83: 2898–2900. 10.1063/1.1614840

8

CHAPTER

POLYHEDRAL SYMMETRY AND QUANTUM MECHANICS

G. S. Anagnostatos

Institute of Nuclear and Particle Physics, National Center for Scientific Research, Demokritos,

ABSTRACT

A thorough study of regular and quasi-regular polyhedra shows that the symmetries of these polyhedra identically describe the quantization of orbital angular momentum, of spin, and of total angular momentum, a fact which permits one to assign quantum states at the vertices of these polyhedra assumed as the average particle positions. Furthermore, if the particles are fermions, their wave function is anti-symmetric and its maxima are identically the same as those of repulsive particles, e.g., on a sphere like the spherical shape of closed shells, which implies equilibrium of these particles having average positions at the aforementioned maxima. Such equilibria on a sphere are solely satisfied at the vertices of regular and quasi-regular polyhedra which can be associated with the most probable forms of shells both in Nuclear Physics and in Atomic Cluster Physics when the constituent atoms possess half integer spins. If the average sizes of the constituent particles are known, then the average sizes of the resulting shells become known as well. This association of Symmetry with Quantum Mechanics leads to many applications and excellent results.

Keywords: Symmetry; Polyhedral Symmetry; Angular Momentum and Polyhedra; Nuclear Structure and Symmetry; Isomorphic Shell Model; Shell Clustering in Nuclei; Binding Energies; Radii

1. INTRODUCTION

Geometry in the form of symmetry has been extensively applied in many areas of physics and chemistry. Specifically, in nuclear physics, the employment of polyhedral symmetry has been employed in the form of several models, particularly in the effort to explain the structure of magic numbers [1-7], that is, to explain the exceptional stability of nuclei with the number of neutrons or protons or both equal to 2, 8, 20, 28, 50, 82, and 126. However, in all of these models, the employment of polyhedra, even though successful for several times, was based mainly on intuition and not in rigorous ways, particularly in relation to Quantum Mechanics, which is the physics of the micro-world. Quantum Mechanics was employed later in these models, and their applicability and accuracy of results were limited as well.

In the present work, the relationship between the symmetry of the polyhedra employed and Quantum Mechanics is the starting point before any effort to explain nuclear phenomena. First, it is proved that the quantization of angular momenta ℓ, s, and j expected for central forces is inherent in the structure of the regular and semi-regular polyhedra employed by the model. No approximation whatsoever is involved in this proof. In other words, the relationship of Polyhedral Symmetry to Quantum Mechanics here is on the level of identity.

Another important feature of the present work which is absent from any other model employing polyhedra [1- 7] is that here the nucleons are not considered point particle, but as particles having finite size as predicted from particle physics for neutrons and protons In addition, we will refer to some significant applications of the model to strengthen our arguments even more.

Overall, the motivation of the present work is to make clear that when one deals with regular and semi-regular polyhedra, Quantum Mechanics is inherently involved. This fact permits us to have a pure quantum mechanical treatment of a problem and at the same time to have a geometrical representation. In particular in nuclear physics, the present work provides a way to unite the two milestone models of the field, the

Independent Particle Shell Model and the Collective Model, under one common assumption.

2. RELATION OF POLYHEDRAL SYMMETRIES TO QUANTUM MECHANICS

2.1. Orbital Angular Momentum

For central forces we have the quantization of direction of the angular momentum as a result of quantization of both the angular momentum itself and its projection on the z-axis. This quantization of direction for the onebody problem leaves the angle φ unspecified, while it specifies the azimuthal angle θ according to the relationship

$$\theta_\ell^m = \cos^{-1}\left[m/\sqrt{\ell(\ell+1)}\right] \tag{1}$$

In the case of the many-body problem, however, it seems that both the angles θ and φ are specified as a result of the restrictions imposed on each particle from the neighbouring particles. When all $2(2\ell+1)$ particles for a certain ℓ value are considered simultaneously, it seems that only one value of φ is permissible for each m value. These angles, as we rigorously show shortly, are related to the regular and semi-regular polyhedra [8].

In **Figure 1** the quantization of directions for $\ell = 1 \div 6$ and all associated m values $(m = -\ell, \cdots, +\ell)$ is shown in relation to the cubic-octahedral symmetry in two rows. The first row refers to $\ell + 1$ values of m, while the second row refers to ℓ values of m with respect to an octahedron, a cube (hexahedron), or a cube-octahedron. The choice of polyhedron each time is made for reasons of clarity, while the z-axis has the same direction. The vectors ℓ forming the appropriate angles pass through char acteristic points of the polyhedra employed marked with a solid circle •. Specifically, these points for $\ell = 1$ are the vertices of an octahedron for both sets of m values (Figures 1(a) and (b)); for $\ell = 2$ the vertices and the shown golden section point of a cube (Figures 1(d) and (e)), respectively); for $\ell = 3$ the middles of the edges of a cube-octahedron for both sets of m (Figures 1(g), (h));

for $\ell=4$ the vertices of an octahedron (**Figure 1**(j)) and the vertices of a cub-octahedron (**Figure 1**(k)); for $\ell=5$ the middle points of the shown line segments of an octahedron (**Figure 1**(m)) and of a cube-octahedron (**Figure 1**(n)); and for ℓ = 6 the third of the shown line segments of a cube (**Figure 1**(p)) and the middle points of the shown line segments of a cub-octahedron (**Figure 1**(q)) [8].

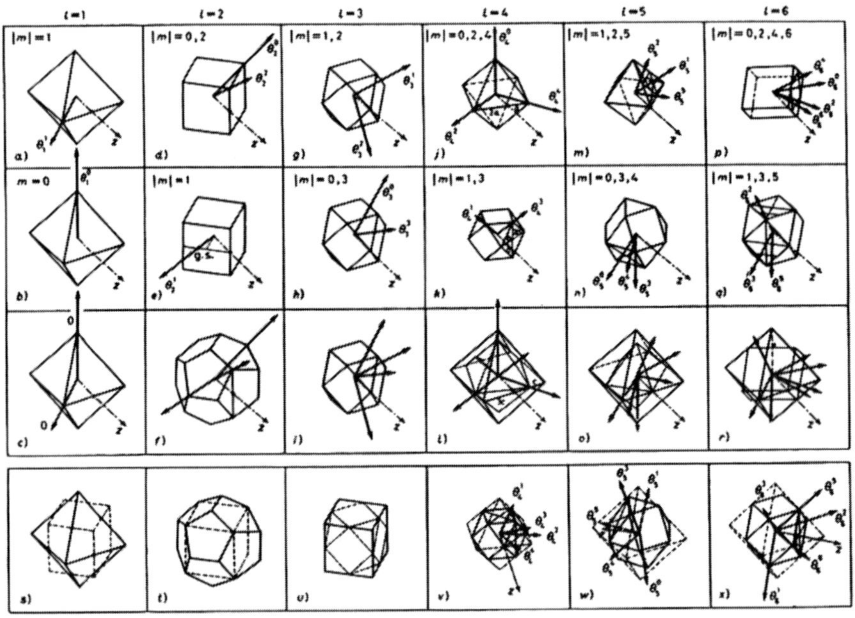

Figure 1. Orbital angular momentum (ℓ) quantization of direction for $\ell=1\div6$ and $m=-\ell,\cdots,\ell$ in relation to the cube-octahedral symmetry. The vectors ℓ form accurately with the z-axis the angle θ_ℓ^m of Eq.1. In row 3 the ℓ vectors for all m of each ℓ value are shown, while in rows 1 and 2 the same vectors are shown in two groups with $\ell+1$ and ℓ values of m, respectively. In all parts of Figures 1(a)-(r) the z-axis has the same direction and all polyhedra shown are oriented with respect to the octahedron of **Figure 1**(a). The vectors ℓ pass through characteristic points of the polyhedra which are marked with a solid circle •, as explained in the text. Figures 1(v), (w), and (x) show an equivalent representation of ℓ vectors shown in Figures 1(k), (n), and (q), respectively and, in addition, two more ℓ vectors (shown with broken lines) already present in Figures l(j), (m), and (p). Figures 1(s)-(u) show the relationships among the polyhedra of the (o) group.

There is an important feature of ℓ vectors. When $\ell = 1 \div 3$, the ℓ vectors of both rows of m (i.e., row 1 and row 2 in **Figure 1**) transform into each other by using a single transformation of the cube-octahedron symmetry group. But as one can see in **Figure 1** when $\ell = 4 \div 6$, the ℓ vectors transform into each other again by a single operation of the cubeoctahedral symmetrygroup but by a different operation for each row of m.

We have demonstrated above that the degenerate states of the same ℓ value for the cases $\ell = 4 \div 6$ have a rotational invariance of Orbital Angular Momentum Quantization of Direction which splits into two sets of m, i.e., one referring to row 1 with $\ell + 1$ values of m and another referring to row 2 with ℓ values of m.

In general, we may argue that the Orbital Angular Momentum Quantization of Direction implies the existence of a fundamental symmetry in nature which could be used as a basis of a physical theory of the structure of matter.

For more details see [8].

2.2. Spin and Total Angular Momentum

In this section the quantization of spin (s) and total (j) angular momenta, in relation to the geometry of regular and semi-regular polyhedra, is presented. That is, in relation to the same polyhedra employed earlier to demonstrate the quantization of the orbital angular momentum (ℓ). These polyhedra are shown in **Figure 2** and are considered superimposed on each other in such a way that they have a common center, a common z-axis, and the most symmetric relative orientation to each other.

In **Figure 3** the quantization of direction of the orbital (ℓ), spin (s), and total (j) angular momenta is shown in parts a), b), and c), respectively, where their relative magnitudes are also shown for $\ell = 1$, $s = 1/2$, and $j = 3/2$. Apparently, the direction of each is unspecified in the polar angle φ, but it is specified in the azimuthal angle θ according to the relationship of Eq.2

$$\theta_{a}^{ma} = \cos^{-1}\left[m_{a} / \sqrt{\alpha(\alpha+1)}\right] \qquad (2)$$

where α stands for ℓ, or s, or j and m_α is the corresponding z-component.

In **Figure 2** the geometry of quantization of the orbital angular momentum is shown for the cases $\ell = 1,2,3,4$ for all m, and $\ell = 5$ for $m_\ell = 5$.

The quantization of direction of the spin, according to Eq.2, is given by the angles

$$\theta_{s=1/2}^{ms=1/2} = 54^\circ 44'08'' \text{ and } \theta_{s=1/2}^{ms=-1/2} = 125^\circ 15'52'' \tag{3}$$

These two angles are identical to characteristic angles of the cube-octahedron (**Figure 2**(c)). The points A, B, C are the middles of the edges of the cube-octahedron. These spin angles are also characteristic angles of the hexa-hedron, dodeca-hedron, octa-hedron and all other polyhedra of**Figure 2** because of their interrelationships shown in Figures 2(e) and (f).

The geometry of the quantization of the total angular momentum is demonstrated in Figures 2(a)-(d). For example consider $\ell = 3(m_\ell = 1)$, $s = 1/2(m_s = 1/2)$, $j = 7/2(m_j = 3/2)$ illustrated in**Figure 2**(c). Eq.2 now gives

$$\theta_{\ell=3}^{m\ell=1} = 73^\circ 13'17'', \theta_{s=1/2}^{ms=1/2} = 54^\circ 44'08'', \theta_{j=7/2}^{mj=3/2} = 67^\circ 47'32'' \tag{4}$$

Both vectors ℓ and s go through the middle of edges of the cube-octa-hedron (the number 1 in $1_{6\text{-}8}$stands for the middle of the edge). Applying the parallelogram rule to these vectors, we find that the resultant vector has a magnitude $\sqrt{7/2 * 9/2}$ and a projection on the z-axis $3/2$, and forms an angle $\theta = 67^\circ 47'32''$ with the z-axis. That is, this resultant vector is identical to. J = ℓ + s with $j = 7/2$ and $m_j = 3/2$.

Other cases examined are apparent in Figures 2(a)-(d). For more details see [9].

3. APPLICATIONS OF SYMMETRIES TO QUANTUM MECHANICS

3.1. Symmetry Description of the Independent Particle Model (IPM)

In **Figure 4**(a) in relation of an octahedron, the quantization of the direction of the orbital angular momentum for $\ell = 1$ is given. All expected directions are axes of symmetry passing through pairs of two opposite

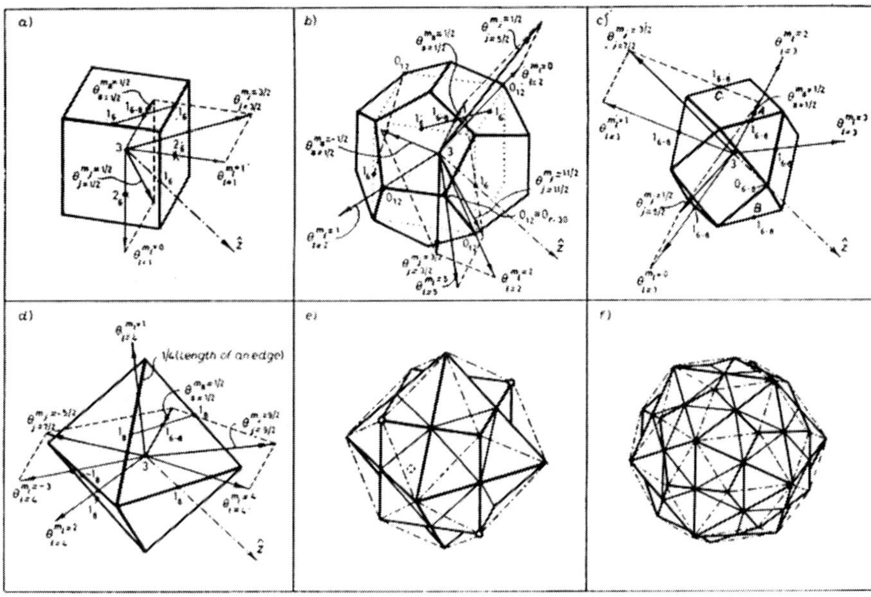

Figure 2. Equilibrium (Leech) polyhedra for the demonstration of the quantization of orbital, intrinsic, and total angular momenta. The numbers 0, 1, 2, 3 stand for the vertex, the middle of an edge, the center of a face, and the center of a polyhedron, respectively. The prefixes 6, 8, 6-8, 12, and r-30 refer to a hexa-hedron (cube) (part a), an octa-hedron (part d), a cube-octahedron (part c), a dodeca-hedron (part b) (where an inscribed cube marked with dotted lines is shown as well), and a rhombic triaconta-hedron, respectively. The symbol θ_ℓ^{m} stands for an orbital-angular-momentum vector ℓ having a projection m_ℓ on the z-axis, which is common for all polyhedra. All the θ vectors shown correspond to the quantization of direction when ℓ is a constant of the motion. The symbol θ_s^{ms} stands for a spin vector s having a projection m_s on the z-axis. When the parallelogram rule is applied to add a $\theta_\ell^{m\ell}$ vector and another θ_s^{ms} vector, for the cases shown, the correct total-angular momentum vector j results (having the correct projection m_j on the z-axis) marked as θ_j^{mj}. Part e): Interconnections among a tetra-hedron (only vertices are shown marked with o), a hexa-hedron, an octa-hedron, a cube-octahedron (dot-dashed lines) and a rhombic dodecahedron. Part f): Interconnection among a dodeca-hedron, an icosa-hodron, an icosi-dodeca-hedron and a rhombic triaconta-hedron (dot-dashed line).

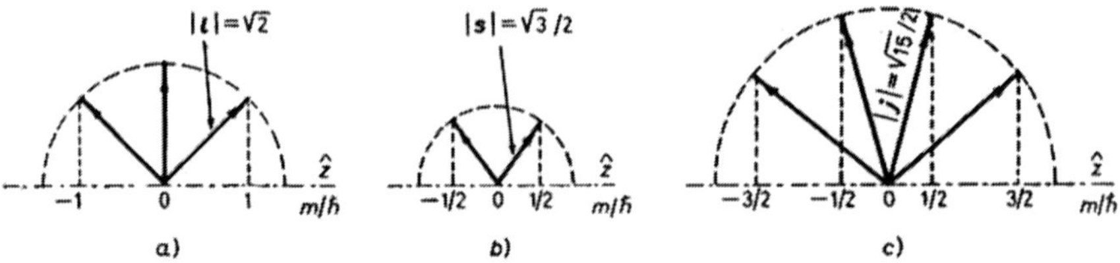

Figure 3. Relationship between the magnitudes of orbital angular momentum a), spin b), and total angular momentum c), and their z-components.

vertices of the octahedron, while in **Figure 4**(b) the quantization of direction for the spin $1/2$ are given in relation to the same polyhedron. The two directions of spin shown pass through the middles of two opposite edges of a cub-octahedron. In Figures 4(c) and (d) the total angular momentum j for $j = 1/2,\ m_j = \pm 1/2$ and $j = 3/2$, $\pm 3/2\ m_j =$, respectively, are derived by applying the parallelogram rule to the orbital angular momentum of **Figure 4**(a) and to the spin of **Figure 4**(b). In **Figure 4**(e), however, the case $j = 3/2$, $m_j = \pm 1/2$ is presented, where neither ℓ nor s come from Figures 4(a) and (b). In Figures 4(c)-(e) j, and not ℓ or s, is the constant of the motion, thus the ℓ and s do not necessarily maintain the same directions in space as in Figures 4(a) and (b). The j in **Figure 4**(e) is determined in relation to the j of Figures 4(c) and (d) according to the law of addition of angular momenta, as will become apparent in a short time. As shown, this is the only case in which j does not pass through the middle of the relevant line segment. For simplicity in Figures 4(c)-(e) the cases of j with negative m_j are omitted. However, j (with $+m_j$) equals −j (with $-m_j$).

So far, in Figures 4(c)-(e), we have determined in relation to an octahedron, the total angular momenta for individual particles which have pairs of j and m_j values appropriate for an assignment to p states. This, of course, guarantees only that the angles of j with the z-axis, i.e. the angles $\theta_j^{m_j}$, are correct and does not guarantee that the vectors j themselves are those expected by the IPM for p states, which is the purpose of this section. The vectors j of Figures 4(c)-(e) must have, in addition, the appropriate relative orientation in such a way that, when the j of two individual particles (e.g. j_1 and j_2) are coupled together, the correct total J of the system results. That is the vectors j_1 and j_2 must form the appropriate angle.

Figures 4(f)-(n) shows that the j vectors of Figures 4(c)-(e), in the framework of the symmetry of an octahedron, possess the appropriate relative orientations which lead to the correct coupling of the total angular momenta of two particles for all cases with $\ell = 1$. Specifically, Figures 4(f) and (g)demonstrate that the j of Figures 4(c)-(e) result in the correct total J for the system of two particles with $j_1 = j_2 = 1/2$. Figures 1(h)-(l) demonstrate that these j also result in the correct J, where $j_1 = j_2 = 3/2$. Figures 4(m)-(n) demonstrate the same when $j_1 = 1/2$ and $j_2 = 3/2$ as registered on each relevant part of the figure. Thus it has been found that the j vectors of Figures 1(c)-(e) are indeed the IPM j.

For more details see [10].

3.2. Applications to Nuclear Structure

3.2.1. Semi-Classical Isomorphic Shell Model

Here, we present the semiclassical part of the model, which has been used many times [11-13] in place of the quantum mechanical part of the model [14], in the spirit of the Ehrenfest theorem [15,16] (that for the average values, the laws of Classical Mechanics are valid). The vertices of the polyhedra of **Figure 5** stand for the distribution of the maxima of the wave function for nucleons due to the anti-symmetric requirement of this function. The Ehrenfest theorem for the observables of position $(\boldsymbol{R})$ and momentum $(\boldsymbol{P})$ takes the form (see all details in [16]).

$$d\langle \boldsymbol{R} \rangle / dt = (1/m) \langle \boldsymbol{P} \rangle \tag{5}$$

And

$$d\langle \boldsymbol{P} \rangle / dt = -\langle \nabla V(\boldsymbol{R}) \rangle \tag{6}$$

For simplicity here, the case of a spinless particle in a scalar stationary potential $V(\boldsymbol{r})$ is considered.

The quantity $\langle \boldsymbol{R} \rangle$ represents a set of three time-dependent numbers $\{\langle X \rangle, \langle Y \rangle, \langle Z \rangle\}$ and the point $\langle \boldsymbol{R} \rangle (t)$ is the center of the wave function at the instant t. The set of those points which correspond to the various values of t constitutes the trajectory followed by the center of the wave packet.

From Eqs.5 and 6 we get

$$m\mathrm{d}^2 \langle \boldsymbol{R} \rangle / \mathrm{d}t^2 = -\langle V(\boldsymbol{R}) \rangle \tag{7}$$

Furthermore, it is known [16] that for special cases of force, e.g., for the harmonic oscillator potential assumed by the model, the following relationship is valid:

$$\langle V(\boldsymbol{R}) \rangle = [V(\boldsymbol{r})]_{\boldsymbol{r}=\langle \boldsymbol{R} \rangle} \quad ,(8)$$

Where

$$-[V(\boldsymbol{r})]_{\boldsymbol{r}=\langle \boldsymbol{R} \rangle} = \boldsymbol{F} \quad .(9)$$

That is, for this potential the average of the force over the whole wave function is rigorously equal to the classical force F at the point where the center of the wave function is considered. Thus, for the special case of potential considered here, the motion of the center of the wave function precisely obeys the laws of classical mechanics [16]. Any difference between the quantum and the classical description of the nucleon motion exclusively depends on the degree that the wave function may be approximated by its center. Any such difference would contribute to deviations between the experimental data and the predictions of the semiclassical part of the model employed.

Thus, in the present semiclassical treatment the nuclear problem is reduced to that of studying the centers of the wave functions of the constituent nucleons or, in other words, of studying the average positions of these nucleons [17].

We further proceed with the help of **Figure 5** which is identical to that figure employed in [11-13], where the most probable forms and average sizes of the first three proton and the first three neutron shells are presented. It is essential to mention that these average sizes solely depend on the average size of a proton, $r_p = 0.860$ fm, and that of a neutron, $r_n = 0.974$ fm. Each occupied vertex configuration of this figure corresponds to a quantum state configuration with definite angular momentum and energy. More details of the figure are given in its caption.

The expressions of the two-body (two Yukava) potential V employed [18], for the present semi-classical treatment, of the kinetic energy T [19], of the spin-orbit energy V_{LS} [20], and of the binding energy E_B are given in Eqs.10-14, respectively. Isospin term in Eq.14 is not needed since the isospin is here taken care of by the different shell structure (forms and sizes) between proton and neutron shells, as apparent from **Figure 5**.

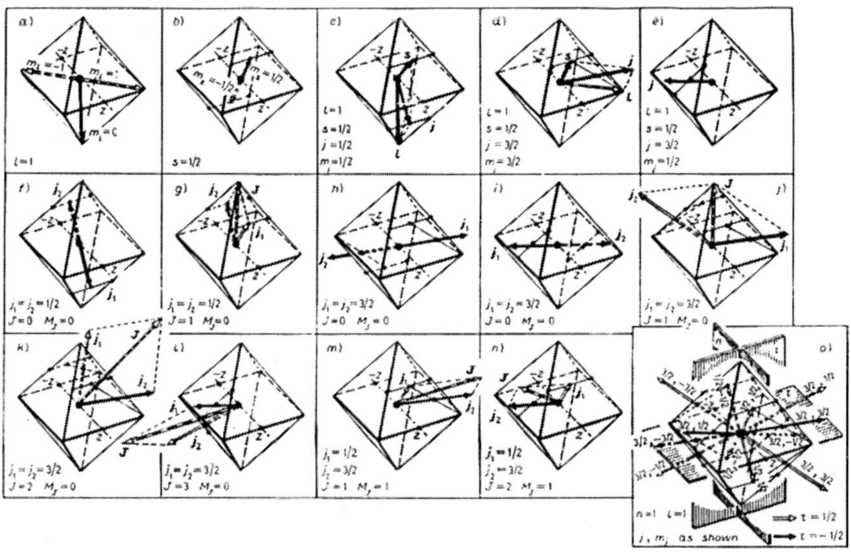

Figure 4. Octa-hedral symmetry description of the Independent Particle Model (IPM) for the 1p-shell. Each mark • stands for the middle point of the corresponding straight-line segment, while large • stands for the center of the octa-hedron. Dark arrows stand for protons and open ones for neutrons, or the opposite, but no mixed [10]. Orbital angular-momentum vectors stand for both protons and neutrons and are presented by dark-open arrows [10]. All j shown are in a relative scale; a) and b) quantisation of direction, when the orbital angular momentum ℓ or spin s are constants of the motion, respectively; c)-e) considered together: quantization of direction when the total angular momentum j is a constant of the motion; f)-n) total angular momentum J of two particles with j_1 and j_2 as registered on each related part; o) assignment of a set of five quantum numbers (n, ℓ, j, m_j and τ) to each j vector of an IPM description of p-shells.

$$V_{ij} = 0.993*10^{17}*e^{-31.2334 r_{ij}}/r_{ij} - 241.193*e^{-1.4534 r_{ij}}/r_{ij} \tag{10}$$

$$\langle T \rangle_{n\ell m} = \left(\hbar^2/2M\right)\left[1/R_{max}^2 + \ell(\ell+1)/\rho_{n\ell m}^2\right] \tag{11}$$

$$\Sigma_i V_{LSi} = \lambda\Sigma_i * (\hbar\omega_i)^2/(\hbar^2/m) * \ell_i s_i \ (\lambda = 0.03, \text{ the third parameter of the model}) \tag{12}$$

$$\hbar\omega_i = (\hbar^2/M)(n+3/2)/\langle r_i^2\rangle \tag{13}$$

$$E_B = \Sigma_{ij} V_{ij} - \Sigma\langle T\rangle_n m - \Sigma_i V_{LSi}, \tag{14}$$

where

- V_{ij} is the potential energy between a pair of nucleons i, j at a distance r_{ij}• n, ℓ, m are the quantum numbers characterizing a polyhedral vertex standing for the average position of a nucleon at the quantum state n, ℓ, m.
- ℓ_i and s_i stand for the orbital angular momentum quantum number ℓ and the intrinsic spin quantum number s of any nucleon i.
- M is the mass of a proton M_p or of a neutron M_n• R_{max} is the outermost proton or neutron polyhedral radius (R) plus the relevant average nucleon radius r_p for a proton and r_n for a neutron, (i.e., R_{max} is the radius of the nuclear volume in which protons or neutrons are confined)• $\rho_{n\ell m}$ is the distance of a nucleon average position at a quantum state (n, ℓ, m) from its orbital angular momentum at the direction $_n\theta_{\ell}^{m}$.

The parameters of the model are the following five: the two-size parameters r_p and r_n, the two parameters from the second term of Eq.10 (since the first term is applicable only for scattering problems), and the one parameter, λ, from Eq.12. With the help of these parameters all quantities R_{max}, $\rho_{n\ell m}$, and $\hbar\omega_i$ in Eqs.10-14 are obtainable by employing the coordinates of the nucleon average positions derived by the information given in Figure 5.

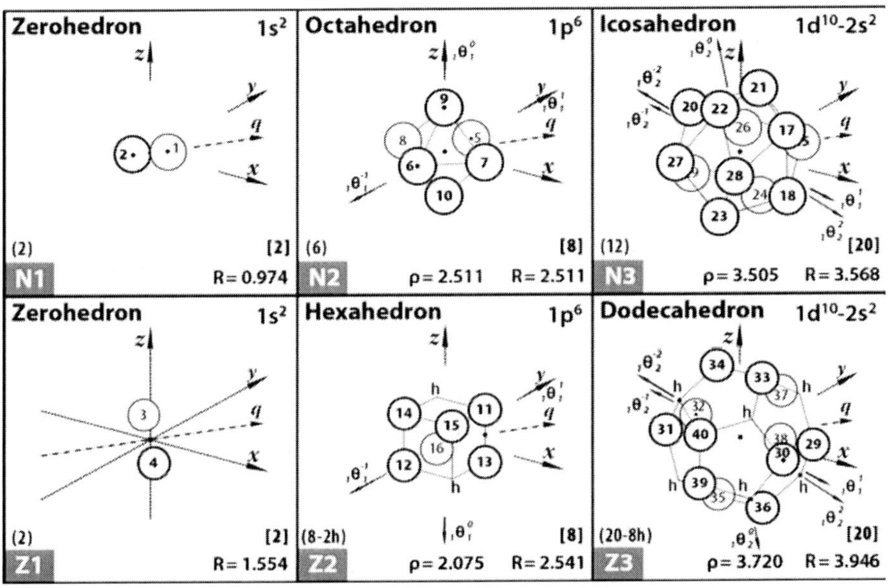

Figure 5. The space of the Isomorphic Shell Model for nuclei up to N = 20 and Z = 20. The equilibrium polyhedra in row 1 (2) stand for the most probable forms and average sizes of the first three neutron (proton) shells. The vertices of polyhedra (numbered as shown) stand for average positions of nucleons in definite quantum states (τ, n, ℓ, m_ℓ, s). Central axes standing for the quantization of directions of the orbital angular momentum are labelled as θ_ℓ^m and pass through the points marked by small solid circles •. At the bottom-left of each block the numbering of this polyhedron proceeded by the letter Z (N) for protons (neutrons) is given. Over this the number of polyhedral vertices and the number of possible unoccupied vertices (holes, h) are also given. At the bottom-right of each block the radius of polyhedron is listed. Over this the cumulative number of vertices of all previous polyhedra and of this polyhedron is also given. Finally, at the bottom-center of each block the distance $\rho_{n\ell m}$ of the nucleon average position nℓm from the relevant axis θ_ℓ^m is given.

In **Table 1** the results of applying the present version of the model for ^4He, ^{12}C, ^{16}O, and ^{40}Ca, together with corresponding experimental energies are given. The good comparisons between model predictions and experimental energies are apparent. Here, calculations of radii are not presented. Such predictions for more nuclei are given in the next section with a full quantum mechanical description of the Isomorphic Shell Model. In the framework of predictions of radii, results are identical for both versions of the model.

Further interesting applications of this version of the model on nuclear structure and reactions are included in Refs. [11-13] and [21,22], respectively.

3.2.2. Quantum Mechanical Isomorphic Shell Model

As is well known the constituent of a nucleus are protons and neutrons which all are fermions, thus their total wave function is anti-symmetric. The locations of maxima of such a wave function on a sphere (like the spherical shape of a complete nuclear shell) are identical to those expected if a repulsive force (of unknown nature) is acting among the constituent fermions (protons and neutrons) [23]. For stability these maxima, under the aforementioned repulsive force, should be in equilibrium.

This problem of equilibria of repulsive particles on a sphere was solved by John Leech in1957 [24]. Such equilibria are obtained only for specific number of particles and are related either to the number of vertices, or to the number of middles of edges, or to the number of middles of faces, or to combinations of numbers of these categories of a regular or semi-regular polyhedron [14].

Such equilibria lead to polyhedral most probable forms which, taken in specific sequence, are presented in **Figure 6** and are related to specific nuclear shells and sub-shells. If further, nucleons of finite sizes are considered (specifically, $r_p = 0.860$ fm and $r_n = 0.974$ fm, which constitute the only size parameters of the model) the average size R of the aforementioned polyhedral shells and subshells are obtained. They are those written at the bottom of each block of the figure.

Apparently, the polyhedra of **Figure 6** present the most probable forms and average sizes of nuclear shells and sub-shells and at the same time are the same polyhedra as the ones of Figures 1-4, i.e., they are again the regular and semi-regular polyhedra. That is, these polyhedra inherently possess quantum mechanics and at the same time they are those polyhedra possessing equilibrium of forces when repulsive particles (or equivalently fermion average positions) fill up their vertices, or middles of edges, or middles of faces.

Some more comments should be made concerning the polyhedra of **Figure 6** and the major models of nuclear structure, i.e., the Shell Model and the Collective Model.

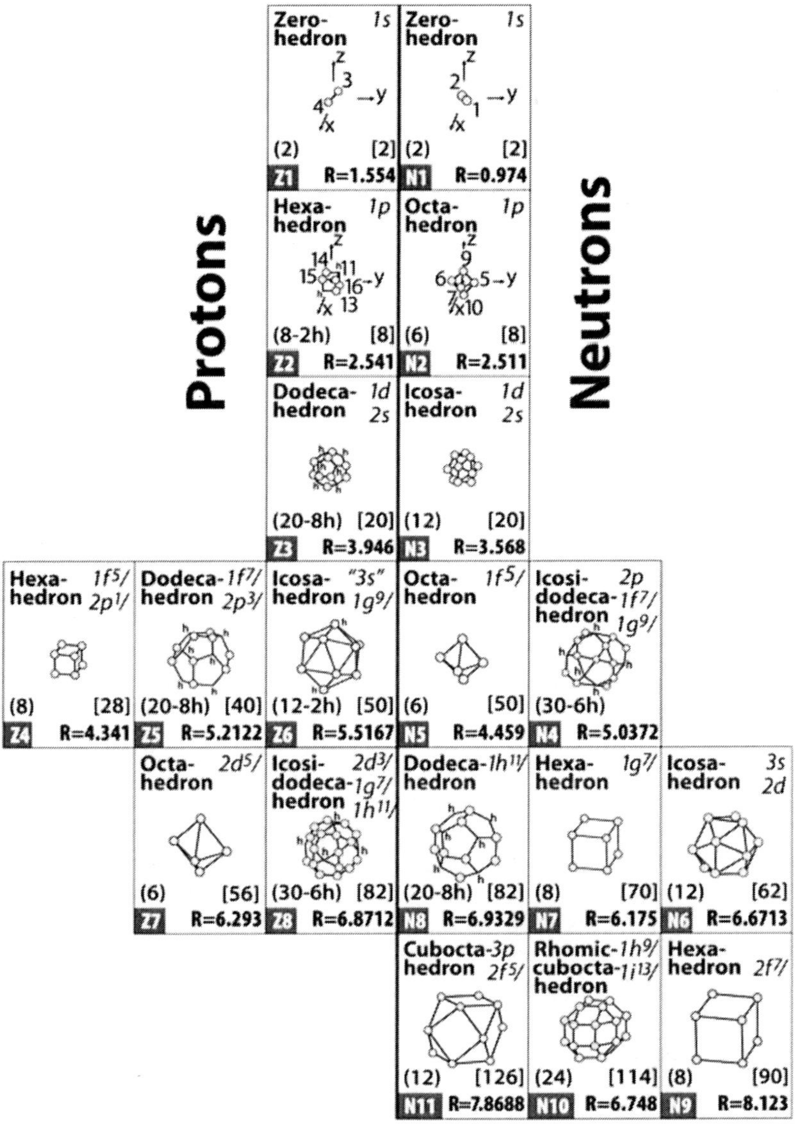

Figure 6. Most probable forms and average sizes of nuclear shells and sub-shells for nuclei up to Z = 82 and N = 126. Other notations are as in the caption of **Figure 5**.

The first model assumes independent particle motion of the constituent particles of a nucleus, which is equivalent to assuming zero forces among these particles in a strong field of strong interactions like a nucleus. In addition, this model it does not give the range of applicability of the assumption and how one can understand the difference of stability (if any) between the magic numbers and the other (about 300) stable nuclei spread in the chart of the nuclides. Moreover, the shell model tries to explain the magic numbers by arbitrarily assuming a strong spin orbit interaction valid only for magic numbers, which is not a necessary assumption for other models (like the present one) to explain these numbers.

Table 1. Binding energies in MeV according to semi-classical isomorphic shell model.

Nuclei	^{4}He	^{12}C	^{16}O (0p-0h)	^{16}O (4p-4h)	^{40}Ca
n	1-2	1-2	1-2	1-2	1-2
		5-8	5-10	5-8	5-10
Occupation			25-27		17-28
p	3-4	3-4	3-4	3-4	3-4
		11-14	11-16	11-14	11-16
			29-31		29-40
$\Sigma_{ij}V_{ij}$	44.61	185.43	262.52	259.93	771.70
$\Sigma\langle T\rangle_{tdm}$	−14.23	−85.32	−124.57	−115.44	−362.97
$\Sigma_i V_{hs}$	0.00	+0.76	1.02		0.00
V_{Coul}	−0.46	−6.75	−12.35	−11.39	−64.81
E_{coll}	−1.70	−1.70	−1.70	−1.70.	−1.70
$E_B$mod.	28.24	92.46	123.90*0.55	+132.42*0.45 = 127.73	342.22
$E_B$exp.	28.30	92.16		127.62	342.06

The second model assumes strong interaction among the constituent particles of a nucleus leading to an average form of the nucleus which can possess collective motion. This assumption apparently contradicts the shell model assumption of zero forces among the constituent particles of a nucleus. The model has many successes through out the periodic table of the elements. However, it cannot predict the moment of inertia of the resulting rotational spectra at the same time. This moment of inertia is empirically derived each time from the rotational spectrum offered by relevant experiments.

Apparently, these two models were milestones at the time when they appeared. Without these two basic models we could not have reached the present level of understanding about the nucleus. After so many years since their appearance, there is an effort to try to go beyond them.

The Isomorphic Shell Model takes advantage of the equilibrium polyhedra to explain the magic numbers without assuming strong spin orbit interaction and at the same time provides the average structure of a nucleus via the average nucleon positions of the constituent nucleons. Thus, it provides the necessary moment of inertia as the moment of inertia of a specific part of the nucleus which rotates around an axis perpendicular to a symmetry axis of this rotating part. Also, the equilibrium of forces obtained when an equilibrium polyhedron is filled is equivalent to the zero forces assumed by the shell model. However, this equilibrium of forces happens not only for magic numbers, but at any time an equilibrium polyhedron is filled up, a fact which explains stable nuclei throughout the periodic chart of nuclides.

In applying the quantum isomorphic shell model a central potential of the following form is applied for the nucleons of each proton or neutron shell (and not for all nucleons in a nucleus):

$$V(r) = v_o + 1/2\, m\omega^2 r^2 \tag{15}$$

where v_0 and ω are different parameters for each proton or neutron shell. Due to the two assumptions below, however, the final number of parameters is substantially reduced.

1) The $\hbar\omega$, for each shell, is determined [20] according to Eq.16,
2)

$$\hbar\omega = \left(\hbar^2/m\right)\left(n+3/2\right)/\left\langle r^2\right\rangle \tag{16}$$

where $\left\langle r^2\right\rangle^{1/2}$ is the average size of a specific proton or neutron shell which remains constant for all nuclei. All these sizes of polyhedra are given in **Figure 6**. The only parameters necessary in determining these sizes are the two size parameters r_p and r_n, as explained earlier.

2) The parameter of the depth of the potential for each proton or neutron shell is determined according to Eq.17

$$E_j = v_j - \hbar\omega_j\left(n_j + 3/2\right) = v_i - \hbar\omega_i\left(n_i + 3/2\right) = E_i$$ or

$$v_j = v_i - \hbar\omega_i\left(n_i + 3/2\right) + \hbar\omega_j\left(n_j + 3/2\right) = V_0 + \hbar\omega_j\left(n_j + 3/2\right) \tag{17}$$

This assumption implies that all nucleons in a nucleus are equally bound in their own potentials (excluding Coulomb and spin-orbit interactions).

It is apparent from Eq.17 that, since according to the previous assumption a) all ħω are already determined as above by applying Eq.16 with respect only to the two size parameters r_p and r_n, the depth v_i of the potential for each shell i can be defined with the knowledge of only one additional parameter V_0 (This is the third universal parameter of the present model equal to 40.268 MeV.)

Solving Schroedinger's equation for the potential of Eq.15, the following general equation for the wave function is obtained [20]:

$$L_k^{\ell+1/2}(z)=\left\{\left[\Gamma(1/2N+1/2\ell+1)\right]^2/k!\Gamma(\ell+3/2)\right\}_1 \cdot F_1(-k;\ell+3/2;z) \quad ,(18)$$

where ${}_1F_1(\alpha;c;z)=1+(\alpha/c)z+\alpha(\alpha+1)z^2/c(c+1)2!$
$+\alpha(\alpha+1)(\alpha+2)z^3/c(c+1)(c+2)3!+\cdots$,

with $\alpha=-k, z=2\alpha r^2$, and $\alpha=m\omega/2\hbar$. The series terminates with the term $(-1)^k\left[\Gamma(c)/\Gamma(c+k)\right]z^k$ and the various quantum numbers involved are $n=0,1,2,\cdots;\ k=0,1,2,\cdots;k\le n/2$, $\ell=0,1,2,\cdots,\ell\le n$, and $\ell=n-2k$ [20].

The explicit forms of equations of the wave functions for a harmonic oscillator potential derived from the recursion formula Eq.18 can be found in [14] for all wave functions of nuclei up to Z = 126 and N = 184.

Due to the fact that in the model ħω is different for the different shells, the wave functions with the same orbital angular momentum quantum number ℓ are not orthogonal. For these wave functions Gram-Smidth's technique is applied [25]. In [14] some relevant guide equations to facilitate this orthogonalization are given.

1) Binding Energies

The binding energy for each quantum state in a harmonic oscillator potential is given by Eq.19:

$$E_B = v - \hbar\omega(n + 3/2) \tag{19}$$

In Eq.19 all ħω come from Eq.16 (with respect to the only two size parameters r_p and r_n) and all potential depths v come from Eq.17 (with respect to the only one additional potential parameter V_0). Thus, E_B from Eq.19 is determined with respect to only 3 universal parameters.

Now, since the nuclear problem basically refers to two-body forces, in order to avoid the double counting, the potential portion of the second part of Eq.19 should be divided by two. Furthermore, since according to the Virial theorem half of the quantity $\hbar\omega(n + 3/2)$ is potential energy and half is kinetic energy, Eq.19 takes the form of Eq.20

$$E_B = 1/2\left[v - 1/2\,\hbar\omega(n + 3/2)\right] \\ - 1/2\,\hbar\omega(n + 3/2) = 1/2\,v - 3/4\,\hbar\omega(n + 3/2) \tag{20}$$

Given that the potential v, according to Eq.17, is

$$v = V_0 + \hbar\omega(n + 3/2) \quad ,(21)$$

the final expression of binding energy for each proton or neutron state takes the form of Eq.22.

$$(E_B)_1 = 1/2\left[V_0 + \hbar\omega(n + 3/2)\right] \\ - 3/4\,\hbar\omega(n + 3/2) = 1/2\,V_0 - 1/4\,\hbar\omega_1(n_1 + 3/2) \quad .(22)$$

The index 1 in Eq.22 refers to all states where the orbital angular momentum quantum number ℓ appears for the first time.

However, for a second, a third and a fourth appearance of a state with the same ℓ, in the place of the quantity $\hbar\omega(n + 3/2)$ in Eq.22 we should take the corresponding quantity due to the necessary orthogonalization by using Eqs.9-11 of [14]

Finally, the total binding energy of a nucleus in the model is given by Eq.23

$$(E_B)_{total} = \sum_{i=1}^{A} (E_B)_i + \sum_{i=1}^{A} V_{LiSi} - \sum_{i \neq j}^{Z} E_{Cij}$$,(23)

where the spin-orbit term is given [20] by Eq.24

$$\sum_{i=1}^{A} V_{LiSii} = \lambda \sum_{i=1}^{A} \left[i/r * \, V(r) / \, r \right] * \ell_i s_i$$
$$= \lambda \sum_{i=1}^{A} (\hbar \omega_i)^2 / \left(\hbar^2 / m \right) * \ell_i s_i$$,(24)

with λ = 0.03 (This is the fourth and the last universal parameter of the present version) and the Coulomb term is given by Eq.25:

$$\sum_{i \neq j}^{Z} E_{Cij} = \sum_{i \neq j}^{Z} e^2 / d_{ij}$$.(25)

The distance d_{ij} between any two proton average positions in **Figure 6** is calculated from the coordinates of the proton polyhedral vertices (standing for the proton average positions) in this figure (see section 2.2 of [14]).

As in the former section, an extra term in Eq.23 due to isospin is not needed since the isospin is here taken care of by the different average shell structure between protons and neutrons as apparent from **Figure 6**.

2) Mean Radii

Due to the way the wave functions have been correlated with the size of nuclear shells via Eq.16, average radii can be calculated by using simple formulas, as seen below.

Average charge radius:

$$\langle r^2 \rangle_{ch} = \sum_{i=1}^{Z} r_i^2 / Z + r_{\text{ch.proton}}^2 - r_{\text{ch.neutron}}^2 * N/Z \quad ,(26)$$

where $r_{ch.proton} = r_p = 0.86$ fm and $r_{ch.neutron} = 0.34$ fm [7].

Average neutron radius:

$$\langle r^2 \rangle_n = \sum_{i=1}^{N} r_i^2 / N + r_{\text{neutron}}^2 \quad ,(27)$$

where $r_{neutron} = r_n = 0.974$ fm .

Average mass radius:

$$\langle r^2 \rangle_m = \left[\sum_{i=1}^{A} r_i^2 + Z * r_p^2 + N * r_n^2 \right] / A \quad .(28)$$

All values of r_i needed are included in **Figure 6** (see right corner at the bottom of each block).

In **Table 2** the binding energies of ^{40}Ca, ^{48}Ca, ^{54}Fe, ^{90}Zr, ^{108}Sn, ^{114}Te, ^{142}Nd, and ^{208}Pb. are listed from [14]. Also for the same nuclei the average charge, neutron, and mass radii are listed again from [14]. For both energies and radii available experimental values are given for comparison. The very good agreements between predictions and experimental values are apparent.

3.2.3. Other Applications

Besides the above given applications of the Isomorphic Shell Model, it is interesting to mention its application to super-heavy nuclei [26], to neutron nuclei [27], to exotic nuclei [28], and its extension to atomic clusters [29], when the constituents atoms have half integer spin and thus can be considered as atomic fermions. Again in these applications the dominant part is the relationship between Symmetry of regular and quasi-regular polyhedra and Quantum Mechanics.

For valuable information concerning geometry, the books [30,31] are very instructive.

4. CONCLUSIONS

In the present work, the regular and semi-regular polyhedra were employed to derive the relationship between their symmetries and quantum mechanics. Particularly, we showed that the symmetries of these polyhedra inherently possess (on the level of identity) the quantization of ℓ, s, and j expected from Quantum Mechanics for central forces. This is a unique property of the present model in comparison with the other models of the nucleus involving polyhedra to explain the magic numbers and other nuclear properties. No relationship of their polyhedra with Quantum Mechanics has been reported. This unique property of the present work to identically connect the symmetry of the regular and semi-regular polyhedra with Quantum Mechanics permits one to assign quantum states at the vertices of these polyhedra considered as the average positions of neutrons and protons in the way explained in the text.

Furthermore, an important feature of the polyhedra employed in the present work (i.e., the regular and semiregular polyhedra) is that they possess the equilibrium property. That is, when repulsive particles (as implied by the Pauli principle for fermions) are assigned at their vertices, standing as their average positions, these particles are at equilibrium whatever the law of force between them may be. In addition, if two sets of fermions are assumed on the same sphere (as neutrons and protons), not only the particles of each set, but also the particles of both sets should be at equilibrium. This equilibrium should be conserved when more polyhedral shells are considered, which implies that the polyhedra should be concentric and in the most symmetric arrangement among themselves. This property of equilibrium of forces is practically equivalent to the assumption of the Shell Model of the nucleus that each nucleon acts as if it is under a zero force.

This equilibrium property leads to the most probable forms of nuclear shells. As mentioned in the introduction, neutrons and protons are not considered as point particles, but as particles with finite size, which have been established from particle physics. The consideration of nucleons with finite size leads to the average size of the nuclear shells and thus of all nuclei. These average positions for the nucleons participating in a collective nuclear rotation form an average shape whose symmetries permit the evaluation of moments of inertia and thus of the energies of the expected rotational bands for this nucleus. This is the way the present work is associated with the Collective Model of the nucleus.

However, there is an important difference between the Collective Model and the present model. In the former, it is assumed that the rotating nucleons strongly interact with each other (like in solid state physics) and form the rotating shape necessary for the rotational band. However, the Collective Model still cannot predict the moment of inertia. The present model does not need this additional assumption. It can predict the moment of inertia and the rotational band based on the average positions of rotating nucleons derived as above.

Table 2. Total binding energies in MeV, average charge, neutron, point neutron-proton, and mass radii.in fm [14].

	^{40}Ca	^{48}Ca	^{54}Fe	^{90}Zr	^{108}Sn	^{114}Te	^{142}Nd	^{208}Pb
$E_B$mod	342.3	416.5	471.3	783.0	915.2	961.1	1185.1	1636.4
$E_B$exp	342.1	416.0	471.8	783.2	914.7	961.2	1185.2	1636.5
$_{ch}\langle r^2\rangle^{1/2}$mod	3.48	3.48	3.72	4.27	4.56	4.61	4.9	5.51
$_{ch}\langle r^2\rangle^{1/2}$exp	3.48	3.48	3.69	4.27	4.56		4.91	5.51
$_n\langle r^2\rangle^{1/2}$mod	3.25	3.65	3.73	4.4	4.7	4.86	5.43	6.18
$[_n\langle r^2\rangle^{1/2} - {}_{ch}\langle r^2\rangle^{1/2}]$mod	−0.29	0.12	−0.03	0.09	0.1	0.22	0.5	0.64
$[_n\langle r^2\rangle^{1/2} - {}_{ch}\langle r^2\rangle^{1/2}]$exp	−0.3	0.12	−0.04	0.09	0.11	-	-	0.55
$_m\langle r^2\rangle^{1/2}$mod	3.38	3.56	3.81	4.36	4.69	4.8	5.22	5.93

A further strong support of this work is provided by its application to specific nuclear properties. First, the ħω values are estimated in both the semi-classical and pure quantum mechanical treatment of nuclei. Specifically, in **Table 1** the nuclei ^{4}He ^{12}C, ^{16}O and ^{40}Ca are treated semi-classically and their binding energies have been estimated with a maximum deviation between predicted and experimental values of 0.30 MeV. In **Table 2** the nuclei ^{40}Ca, ^{48}Ca, ^{54}Fe, ^{90}Zr,^{108}Sn, ^{114}Te, ^{142}Nd, and ^{208}Pb are treated purely quantum mechanically. Properties examined there are the binding energies and radii. The maximum deviation between predicted and experimental values of binding energies is 0.5 MeV. For the average values of charge radii, predicted and experimental values are identical, with the only exception for ^{54}Fe where the difference is 0.03 fm. For the difference between neutron and proton average radii in the same table the deviation is 0.01 fm, except for ^{208}Pb, where the deviation is 0.09 fm and is considered significant. For the average neutron radii, there are no experimental values for comparison.

Other applications of the present model are given elsewhere [14]. Hence, a very strong connection between Polyhedral Symmetry and Quantum Mechanics is provided throughout this work.

Perhaps it is interesting to finish with Albert Einstein's words: Geometry is always the solution. The question is which geometry each time.

REFERENCES

1. Pauling, L. (1990) Regularities in the sequences of the number of nucleons in the revolving clusters for the ground-state energy bands of the even-even nuclei with neutron number equal to or greater than 126. Proceedings of the National Academy of Science USA, 87, 4435-4438, and references therein.

2. Cook, N.D. (1994) Nuclear binding energies in lattice models. Journal of Physics G (Nuclear and Particle Physics), 20, 1907-1917 and references therein.

3. Rae, W.D.M. and Zhang, J. (1994) Triangular alpha cluster geometries and harmonic oscillator shell structure. Modern Physics Letters A, 9, 599-607 and references therein.

4. Robson, D. (1978) Many-body interactions from quark exchanges and the tetrahedral crystal structure of nuclei. Nuclear Physics A, 308, 381-428 and references therein.

5. Naher, U., Zimmermann, U. and Martin, T.P., (1993) Geometrical shell structure of clusters. Journal of Chemical Physics, 99, 2256-2260.

6. Hecht, L. (1988) The geometrical basis for the periodicity of the elements.

7. White, (1987) New hypothesis shows geometry of atomic nucleus. Executive Intelligence Review, 14, 18.

8. Anagnostatos, G.S., Giapitzakis, J. and Kyritsis, A. (1981) Rotational invariance of orbital-angular-momentum quantization of direction for degenerate states. Lettere Nuovo Cimento, 32, 332-335.

9. Anagnostatos, G.S. (1980) The geometry of the quantization of angular momentum (ℓ, s, j) in fields of central symmetry. Lettere Nuovo Cimento, 28, 573-576.

10. Anagnostatos, G.S. (1980) Symmetry description of the independent particle model. Lettere Nuovo Cimento, 29, 188-192.

11. Anagnostatos, G.S., Ginis, P. and Giapitzakis, J. (1998) α-Planar states in ^{28}Si". Physical Review C, 58, 3305- 3315.

12. Kakanis, P.K. and Anagnostatos, G.S. (1996) Persisting α-planar structure in ^{20}Ne. Physical Review C, 54, 2996- 3013.

13. Anagnostatos, G.S. (1999) Alpha-chain states in ^{12}C". Physical Review C, 51, 152-159.

14. Anagnostatos, G.S. (2013) Quantum isomorphic shell model: Multi harmonic shell clustering in nuclei. Journal of Modern Physics, 4, 54-65.

15. Merzbacher, E. (1961) Quantum mechanics. John Wiley and Sons Ltd., New York, p. 42.

16. Cohen-Tannoudji, C., Diu, B. and Laloe, F. (1977) Quantum mechanics. John Wiley & Sons Ltd., New York, p. 240.

17. Anagnostatos, G.S., Antonov, A.N., Ginis, P., et. al. (1998) Nucleon momentum and density distributions in ^{4}He considering internal rotation. Physical Review C, 58, 2115- 2119.

18. Anagnostatos, G.S. and Panos C.N. (1982) Effective twonucleon potential for high-energy heavy-ion collisions. Physical Review C, 26, 260-264.

19. Panos, C.N. and Anagnostatos, G.S. (1982) Comments on a relation between average kinetic energy and meansquare radius in nuclei. Journal of Physics G: Nuclear Physics, 8, 1651-1658.

20. Hornyak, W.F. (1975) Nuclear structure. Academic, New York, pp. 13, 240, 237.

21. Anagnostatos, G.S. (1989) Classical equations-of-motion model for high-energy heavy-ion collisions. Physical Review C, 39, 877-883.

22. Anagnostatos, G.S. and Panos, C.N. (1990) Semiclassical simulation of finite nuclei. Physical Review C, 42, 961- 965.

23. Sherwin, C.W. (1959) Introduction to quantum mechanics. Holt, Rinehart and Winston, New York, p. 205.

24. Leech, J. (1957) Equilibrium of sets of particles on a sphere. Mathematical Gazette, 41, 81-90.

25. Vergados, J.D. (1970) Mathematical methods in physics. Akourastos Giannis.

26. Anagnostatos, G.S. (2008) A new look at super-heavy nuclei. International Journal of Modern Physics B, 22, 4511- 4523.

27. Anagnostatos, G.S. (2008) On the possible stability of tetraneutrons and hexaneutrons. International Journal of Modern Physics E, 17, 1557-1575.

28. Paschalis, S. and Anagnostatos, G.S. (2013) Ground state of ^{4-7}H considered internal collective rotation. Journal of Modern Physics, 4, 66-77.

29. Anagnostatos, G.S. (1991) Fermion/boson classification in micro-clusters. Physics Letters A, 157, 65-72.

30. Cundy, H.M. and Rollett (1961) Mathematical models. 2nd Edition, Oxford University Press, Oxford.

31. Coxeter, H.S.M. (1963) Regular polytopes. 2nd Edition, The Mcmillan Company, New York.

9

CHAPTER

THE MEASUREMENT PROBLEM FROM THE PERSPECTIVE OF AN INFORMATION-THEORETIC INTERPRETATION OF QUANTUM MECHANICS

Jeffrey Bub [1,2,3]

[1]*Philosophy Department, University of Maryland, College Park 20742, MD, USA*

[2]*Joint Center for Quantum Information and Computer Science, University of Maryland, College Park 20742, MD, USA*

[3]*Institute for Physical Science and Technology, University of Maryland, College Park 20742, MD, USA*

ABSTRACT

The aim of this paper is to consider the consequences of an information-theoretic interpretation of quantum mechanics for the measurement problem. The motivating idea of the interpretation is that the relation between quantum mechanics and the structure of information is analogous to the relation between special relativity and the structure of space-time. Insofar as quantum mechanics deals with a class of probabilistic correlations that includes correlations structurally different from classical correlations, the theory is about the structure of information: the possibilities for representing, manipulating, and communicating information in a genuinely indeterministic quantum world in which measurement outcomes are intrinsically random are different than we thought. Part of the measurement problem is deflated as a pseudo-problem on this view, and the theory has the resources to deal with the remaining part, given certain idealizations in the treatment of macrosystems.

Keywords: quantum information; measurement problem; information-theoretic interpretation

1. INTRODUCTION

The salient feature of quantum phenomena that resists any classical explanation is the existence of nonlocal correlations associated with entangled quantum states. Schrödinger ([1], p.555) identified entanglement as "*the* characteristic trait of quantum mechanics, the one that enforces its entire departure from classical lines of thought".

For two quantum systems in a pure entangled state, Colbeck and Renner [2,3] have shown that there can't be a variable λ, associated with the history of the systems before the preparation of the entangled state in the reference frame of any inertial observer, that provides information about the outcomes of measurements on the systems, so that the marginal probabilities conditional on λ for measurement outcomes on the separate systems differ from the quantum probabilities. In other words, the information in a pure entangled state about the probabilities of measurement outcomes is as complete as it could be, which means that the measurement outcomes are intrinsically random events in the sense that they are uncorrelated with any prior events.

Gleason ([4], p. 886) used the term "intertwined" to describe the intricate meshing of commuting and noncommuting quantum observables in a seminal paper in which he proved that it follows from this feature that there is essentially only one way to define probabilities on the structure of quantum properties defined by the two-valued observables in a Hilbert space of three or more dimensions. Intertwinement imposes objective pre-dynamic probabilistic constraints on correlations between events in a similar sense to which Minkowski space-time imposes kinematic constraints on events. The probabilistic constraints of the correlational structure provide the framework for the physics of a genuinely indeterministic universe. They characterize the structure of information for correlations that include the nonlocal correlations of entangled quantum states, which can only occur between intrinsically random events.

An information-theoretic interpretation of quantum mechanics in the sense I have in mind begins with the idea that the relation between quantum mechanics and the structure of information is analogous to the relation between special relativity and the structure of space-time. Insofar as quantum mechanics deals with a class of probabilistic correlations that includes correlations structurally different from correlations that arise in classical theories, the theory is about the structure of information.

Shannon's classical theory of information [5] was about the communication of messages as signals via a communication channel. The fundamental question for Shannon was how to quantify the minimal physical resources required to represent messages produced by a source, so that they could be communicated via a channel and reconstructed by a receiver. For this problem, and related communication problems, the meaning of the message is irrelevant, as Shannon pointed out:

The fundamental problem of communication is that of reproducing at one point either exactly or approximately a message selected at another point. Frequently the messages have *meaning*; that is they refer to or are correlated according to some system with certain physical or conceptual entities. These semantic aspects of communication are irrelevant to the engineering problem. The significant aspect is that the actual message is one *selected from a set* of possible messages. The system must be designed to operate for each possible selection, not just the one which will actually be chosen since this is unknown at the time of design.

—Claude E. Shannon ([5], p. 379)

The classical theory of information was about the "engineering problem" of communicating messages over a channel efficiently. In this sense, the concept of information has nothing to with anyone's knowledge and everything to do with the stochastic or probabilistic process that generates the messages. What we have discovered is that the 'engineering' possibilities for representing, manipulating, and communicating information in a quantum world are different than we thought, irrespective of what the information is about.

The purpose of this paper is to consider the consequences of this view for the measurement problem of quantum mechanics.

2. SOME CLARIFICATIONS

To say that quantum mechanics is about the structure of information is not to endorse a view that "our reality is ultimately made up of information" and that "the laws of Nature are information about information" ([6], pp. 13, 215, 218). The idea that the basic building blocks of reality might be information is intriguing, but I have no idea how to make sense of this. It can't be information in the sense that Shannon had in mind.

Harvey Brown and Chris Timpson [7] have criticized my earlier remarks [8,9,10] along these lines, particularly with reference to the Clifton–Bub–Halvorson result [11] that the observables and state space of a theory formulated in C∗-algebraic terms must be quantum mechanical if the theory satisfies certain information-theoretic constraints. Briefly, the critique focuses on my reference to Einstein's distinction between principle and constructive theories, and the claim that quantum mechanics, like special relativity, should be understood as a principle theory, where the principles concern space-time structure in the case of relativity, and information-theoretic structure in the case of quantum mechanics:

Even more significantly for our purposes, at the end of their paper CBH suggested an analogy between their characterization of quantum mechanics and Albert Einstein's special theory of relativity (henceforth SR). The "foundational significance" of the CBH derivation is, according to these authors, that quantum mechanics should be interpreted as a principle theory, in the sense of the term that Einstein used to describe his 1905 formulation of SR. CBH saw their constraints as analogous to the principles—the relativity principle and the light postulate—used by Einstein to derive the nature of relativistic kinematics. ...Einstein's principle theory route was based on a policy of despair, and represented a strategic retreat from the more desirable but, in his view, temporarily unavailable constructive approach.

—Harvey Brown and Chris Timpson ([7], p. 3)

The principle versus constructive distinction is a bit of a red herring here, as is Einstein's view about the sense in which a principle theory could provide an understanding of physical processes. Later in the paper, Brown and Timpson say (my italics):

Like the above-mentioned ether theorists [Lorentz, Larmor, and Poincaré], Einstein realized that the covariance of Maxwell's equations—the form invariance of the equations—is achieved when the relevant coordinate transformations take a very special form, but Einstein was unique in his understanding that *these transformations, properly understood, encode new predictions as to the behaviour of rigid bodies and clocks in motion.* That is why, in Einstein's mind, *a new understanding of space and time themselves was in the offing.*

—Harvey Brown and Chris Timpson ([7], p. 6)

Quantum mechanics was born with Heisenberg's seminal 1925 "Umdeutung" paper [12] "On the quantum-theoretical re-interpretation [Umdeutung] of kinematical and mechanical relations". Heisenberg thought that the discrete orbits of Bohr's atomic theory were an artificial theoretical fix that "saved the appearances", but this was not the right way to think about the structural features of atoms responsible for the emission and absorption spectra of gases. His proposal was to *re-interpret*classical mechanical quantities, like position, momentum, energy, angular momentum, as operations, later represented by operators that act on and transform the states of quantum systems. The aim was to explain the discrete frequencies of light emitted by atoms without appealing to electron orbits, and later to explain other phenomena that couldn't be explained by classical physics. In a 1925 letter to Wolfgang Pauli, Heisenberg wrote:

All of my meagre efforts go toward killing off and suitably replacing the concept of the orbital paths that one cannot observe.

—David C. Cassidy ([13], p. 197)

In the same year, Max Born and Pascual Jordan published the first formulation of quantum mechanics in the first part of a two-part paper "Zur Quantenmechanik" [14]. Part II of this paper, referred to by historians as the "three-man paper", was co-authored with Heisenberg and published a year later [15].

A noncommutive mechanics with the sort of intertwinement of commuting and noncommuting observables characteristic of quantum mechanics noted by Gleason *"encodes new predictions as to correlations between measurement outcome events,"* which is why *"a new understanding of the structure of information is in the offing,"* analogous to the new relativistic understanding of space and time required by the way rigid bodies and clocks in motion behave. This is the information-theoretic interpretation of quantum mechanics in the sense I have in mind.

3. CONSEQUENCES FOR THE MEASUREMENT PROBLEM

Itamar Pitowsky distinguishes a "big" measurement problem from a "small" measurement problem [16,17]. The "big" measurement problem is the problem of explaining how the dynamics of a quantum measurement process produces a definite outcome. The "small" measurement problem is the problem of explaining how a classical probability distribution over macroscopic measurement outcomes emerges dynamically in a measurement process.

On the information-theoretic interpretation, the "big" measurement problem is a pseudo-problem. If the universe is genuinely indeterministic and measurement outcomes are intrinsically random, then it isn't possible to provide a dynamical explanation of how a system produces a definite outcome when it's measured—that's what it means for the measurement outcomes to be *intrinsically* random. The random selection of a definite outcome in a quantum measurement process is a feature of the nonclassical structure of information in a quantum world, just as Lorentz contraction is a feature of the Minkowski structure of space-time in a relativistic world.

In a classical theory, probabilistic correlations—consider correlations between two systems for simplicity—have the structure of a simplex (a line with two end points or vertices, a triangle with three vertices, a tetrahedron with four vertices, and so on). A point in the interior of the simplex represents an array of joint probabilities for all possible measurements and all possible measurement outcomes on the systems. The vertices of the simplex represent deterministic 0,1 probabilities. A simplex is the only convex polytope for which there is a *unique* representation of any interior point as a convex combination of the vertices. So classical theories are rather special: there is a unique representation of the associated joint probabilities as probability measures over deterministic states. Correlations that can be simulated with local or classical resources form a convex polytope, the local correlation polytope, which is not in general a simplex. A simulation corresponds to a causal explanation of the correlations in a simplex theory. The local correlation polytope is included in a convex set with a continuous boundary representing quantum correlations, and this convex set is in turn included in the set of correlations that satisfy a "no signaling" constraint, a convex polytope (also not a simplex) with vertices representing Popescu-Rohrlich (PR) boxes [18]. In a classical simplex theory where all quantities are definite, it is possible to give a dynamical account of how a particular measurement outcome is produced in a measurement process: the (pure) state of the system undergoes a deterministic reversible dynamical transition from one vertex of the simplex to a vertex associated with the measurement outcome. For correlations that can't be simulated in a simplex theory, there is no analogous dynamical account (if you're not a Bohmian or an Everettian—see below). Any such account would be inconsistent with the "no signaling" principle.

Here's a way to see this. The state of a quantum system can be specified by the probabilities of the outcomes of measurements on some "informationally complete" or "fiducial" set of observables. For a qubit, the observables X,Y,Zform an informationally complete set. Since these observables don't commute, one would need to measure X,Y,Z on many qubits, all in the same quantum state, to estimate the state from the statistics of the measurement outcomes. Suppose Alice sends Bob a qubit in an unknown quantum state and Bob measures the observable *X*. If there were a deterministic reversible dynamics that described how Bob's measurement outcome came about, Bob could reverse the dynamical evolution to recover the original state after a measurement. Bob could measure the observables X,Y,Z as many times as he liked on the *same* qubit and identify the unknown state to any arbitrary accuracy from the statistics of the measurement outcomes.

Now suppose that Alice and Bob share an entangled state, say the Bell state $|\phi^+\rangle = \frac{1}{\sqrt{2}}(|0\rangle|0\rangle + |1\rangle|1\rangle) = \frac{1}{\sqrt{2}}(|+\rangle|+\rangle + |-\rangle|-\rangle)$, where $|+\rangle = \frac{1}{\sqrt{2}}(|0\rangle + |1\rangle), |-\rangle = \frac{1}{\sqrt{2}}(|0\rangle - |1\rangle)$. Suppose Alice chooses to measure either the observable *Z* associated with the basis |0⟩,|1⟩, or the observable *X* associated with the basis |+⟩,|−⟩. If she measures *Z*, her qubit ends up in the state |0⟩ or |1⟩ with equal probability. If she measures *X*, her qubit ends up in the state |+⟩ or |−⟩ with equal probability. The state of Bob's qubit is correlated with the state of Alice's qubit, so after Alice's measurement Bob's qubit will be in an equal weight mixture of the states |0⟩,|1⟩, or in an equal weight mixture of the states |+⟩,|−⟩. If Bob could identify the quantum state of the single qubit in his possession, he could distinguish these two mixtures. Alice could signal to Bob, effectively instantaneously, because Alice and Bob could be any distance apart. So a dynamical account of how a definite measurement outcome is selected would be inconsistent with the 'no signaling' principle. It follows that there can't be a deterministic reversible dynamics that describes how Bob's measurement outcome is produced in a measurement process. If Bob gains information about an observable in an informationally complete set in a measurement, there must be a corresponding loss of information about other observables in the informationally complete set. The irreducible information loss on measurement is a generic feature of any no-signaling theoretical explanation of correlations that lie outside the local correlation polytope.

A quantum "measurement" is a bit of a misnomer and not really the same sort of thing as the measurement of a property of a classical system, like the temperature of a liquid, or the length of a table top. It involves putting a quantum system like a photon in a situation, say a beamsplitter or an analyzing filter, where the photon has to make an intrinsically random transition to one of two macroscopically distinct alternatives recorded in a device like a photon detector. The quantum probabilities for the values of a polarization observable in a quantum state can't be understood in the classical sense as quantifying ignorance about a particular pre-measurement value of the polarization, or ignorance about an "instruction set" that tells a quantum system what value to produce in a measurement of the polarization. If you could understand quantum probabilities in this way, then you wouldn't get the correlations of entangled states.

So quantum probabilities are understood as probabilities of "what you'll obtain if you measure". This is satisfactory only if a measurement is a process in which probabilities of "what you'll obtain if you measure" become probabilities that can be interpreted in the classical sense as measures of ignorance over several possibilities, one of which actually happens. Otherwise, even after a measurement, you still have probabilities of "what you'll obtain if you measure". What you want is to be able to drop the "if you measure" qualification after a measurement with a macroscopic measuring instrument and recording device, and interpret "what you'll obtain" as "what's there". As Einstein put it ([19], p. 907), you want the moon to be there, and not only there when you look (measure). Is this justified in a quantum universe, and if so, how? This is the "small" measurement problem.

The position of a macroscopic pointer, or Schrödinger's cat being alive or dead [20], is a collective property of an enormous number of microsystems making up the pointer or the cat. A quantum model of a macrosystem treats a macrosystem as composed of a potentially infinite number of microsystems. Macro-observables associated with collective properties can be defined for such an idealized macrosystem, and it turns out that, in the limit, as the number of constituent microsystems goes to infinity, the macro-observables all commute with each other, and with all other observables associated with a finite number of microsystems. For a microsystem, the only observables that commute with all other observables are the zero or null observable and the unit or identity observable, which form the "center" of the noncommutative algebra of observables. For a macrosystem, the center of the algebra contains the null observable and the identity observable, as well as all the macro-observables. Since the macro-observables are in the center of the algebra and commute with all observables, they can be associated with definite values, like the observables of a classical system. Different macrostates correspond to different values of the macro-observables. Technically, the state space (Hilbert space) of a macrosystem breaks up into orthogonal "superselection sectors", and the different values of a macro-observable correspond to these superselection sectors. The existence of superselection sectors is associated with a superselection rule that limits pure state superpositions to states in a particular superselection sector, so there are no superpositions of macrostates. See, for example, [21,22].

If a quantum system like a photon is measured by a macroscopic measuring and recording device, you end up with a mixed state, with probabilities over the different possible measurement outcomes (each of which corresponds to a superselection sector) that can be interpreted as classical probabilities in the ignorance sense. In other words, these probabilities are probabilities of "what's there" rather than "what you'll obtain if you measure". The measurement outcomes are definite because the macroscopic measuring and recording device, as far as its macro-observables are concerned, is a classical system, with a commutative algebra of observables, which is equivalent to saying that the algebra of macro-properties is a Boolean algebra. (Geometrically, a Boolean algebra corresponds to a simplex: there is a one-one correspondence between the vertices, edges, and faces of a simplex and the elements of a Boolean algebra. The vertices correspond to Boolean atoms, and the edges and faces correspond to compound Boolean elements.) Putting it in terms of Schrödinger's cat, the cat doesn't end up in an entangled state, where the probabilities defined by the coefficients of the entangled state represent probabilities of "what you'll obtain if you measure" the cat, that is, if you open the chamber and look. Rather, the atom–plus–device–plus–cat ends up in a mixed state in which the cat is either definitely alive or definitely dead, with the appropriate probabilities, interpreted in the usual classical sense as representing ignorance about the two alternatives, one of which actually happens.

This is not to deny that a macroscopic system like a cat could, in principle, exhibit quantum effects of interference or entanglement if it's the object of a quantum measurement with some other system as the measuring instrument, as in recent experiments designed to show that relatively large systems like optomechanical oscillators can be in quantum superpositions or entangled states [23]. In Schrödinger's thought experiment, the cat is treated as the ultimate measuring instrument that registers the decay of the radioactive atom. On the information-theoretic interpretation, the system that plays this role, in Schrödinger's experiment or in an interference experiment with an optomechanical oscillator, must be capable of being modeled as a system with commuting macro-observables, so the macro-properties form a Boolean algebra.

Here's a rough outline of how this works in an idealized "toy model" by Klaus Hepp [24]. In Hepp's model, a macroscopic measurement "pointer" is represented as an infinite array of qubits. The states of the infinite array in which all but a finite number of qubits are in the state $|0\rangle$ belong to one superselection sector, and the states of the infinite array in which all but a finite number of qubits are in the state $|1\rangle$ belong to a second superselection sector. These two superselection sectors represent two macrostates of the pointer, associated with the pointer readings 0 and 1. Initially, the infinite array representing the macroscopic pointer is in the state with all qubits in the state $|0\rangle$, so the pointer reads 0. The measured system is a single qubit, Q. If Q is in the state $|0\rangle$, the array qubit states are unchanged, but if Q is in the state $|1\rangle$, the measurement interaction flips the array qubit states from $|0\rangle$ to $|1\rangle$ in order, as time goes on. If Q is in a superposition of the states $|0\rangle$ and $|1\rangle$, you end up after the measurement interaction with a mixed state over the two pointer macrostates, in which the 0 pointer reading state is correlated with the state $|0\rangle$ of Q, and the 1 pointer reading state is correlated with the state $|1\rangle$ of Q. That's what you want: at the end of the measurement, the measured system Q and the measuring instrument end up in a mixed state, which can be interpreted as a classical probability distribution over definite measurement outcomes. The probabilities of the initial state of Q, probabilities of "what you'll obtain if you measure", become classical probabilities in the ignorance sense over the two possible measurement outcomes, the definite pointer reading states 0 and 1, which are correlated with the states $|0\rangle$ and $|1\rangle$ of Q (the states corresponding to the two possible values of the measured observable).

Now, as Bell pointed out in a critique of Hepp's model [25], this only occurs at the infinite limit, and in Hepp's model the idealization involves not only considering the limit for an infinite number of qubits, but also the infinite time limit—at any finite time after the measurement interaction begins, only a finite number of qubits in the array are flipped, and so an initial pure state in a superselection sector remains in that superselection sector.

What do these infinite limits mean physically, in particular the infinite time limit? Landsman points out ([21], p. 514) that a measurement is comparable to a scattering process and, for measurement as for decoherence, although the process approaches the behavior at the limit very rapidly, there's no finite time at which the process is over:

> Indeed, what would it mean for scattering to be over after some finite time? Which time? As we shall see ...the theory of decoherence requires the limit $t \rightarrow \infty$ as well, and largely for the same mathematical reasons. There as well as in Hepp's approach, the limiting behavior actually tends to be approached very quickly (on the pertinent time scale), and one needs to let $t \rightarrow \infty$ merely to make terms ~exp−γt (with $\gamma > 0$) zero rather than just very small.
>
> —Nicolaas P. Landsman ([21], p. 514)

Similar idealizations are typical in classical statistical physics. A macroscopic classical system has "emergent" properties that are not properties of its constituent microsystems. For example, a liquid can freeze to a solid when cooled and evaporate to a gas when heated, but the solid–liquid–gas phases of a substance are not well-defined properties of individual molecules or finite collections of molecules, and neither is the temperature. Being a solid or a liquid or a gas, or having a certain temperature, is a collective property of enormous numbers of molecules. In classical physics, collective properties like this, which are insensitive to adding or removing any finite number of molecules, are described theoretically in terms of macro-observables associated with potentially infinite numbers of molecules, and the theoretical analysis necessarily involves idealizations. A glass of water has roughly 1024 molecules, but classical statistical mechanics produces a phase transition from a liquid to a gas, or a liquid to a solid, only if the system is idealized as an infinite number of molecules. The phase transition occurs at the infinite limit, but not for any finite number of molecules.

4. ALTERNATIVES

The standard alternatives to this information-theoretic interpretation are Bohm's theory [26,27] or the Everett interpretation [28,29]. Both treat indefiniteness at the quantum level and the intrinsic randomness of quantum events as only apparent, but the devil is in the details.

In Bohm's theory, the values of hidden variables provide missing information that resolves the indefiniteness at the microlevel and, together with the dynamics, explains how measurement outcomes come about, and the correlations between measurement outcomes at the macrolevel. That's the general idea, but the picture is not as straightforward as it might seem at first sight. There are cases where events are registered as occurring when the particle trajectories in Bohm's theory, which are supposed to explain the occurrence of these events, are located somewhere else [30,31,32]. See, for example, the critique by Reinhard Werner that Bohm's theory doesn't really do what it's claimed to do [33].

The Everett interpretation is fundamentally a proposal to solve the "big" measurement problem: a measurement produces a definite outcome in each of many worlds that emerge through the dynamics of decoherence. The basic idea is that the universal quantum wave function is all there is, and each component or "branch" of the wave function evolves separately and represents a complete world as we experience it. Everettians see this as really just standard quantum mechanics, not one interpretation among rivals, but there are some problems and loose ends that complicate the Everettian picture.

The representation of a wave function as a specific superposition of components or branches depends on the basis in the quantum state space, the "coordinate system" with respect to which the wave is represented as superposition. Something physical should determine the basis, since the worlds associated with the branches of the universal wave function depend on the basis. The modern version of the Everett interpretation [34,35] proposes that the basis is selected by decoherence, a feature of the interaction between a macrosystem and its environment. The picture is that a multiplicity of worlds "emerges" as a feature of the universal wave function through decoherence as a dynamical process, which occurs spontaneously with respect to a position basis. In this "multiverse", we ourselves are split into different versions with different experiences associated with the worlds on the different branches of the universal wave function. In particular, the experiences of different outcomes of a quantum measurement occur in worlds on different branches.

The idea is ingenious but the full story is a lot more complicated than this summary suggests. What you get from decoherence is a basis of approximately localized positions, and the worlds that "emerge" in this dynamical process are only well-defined in an idealized sense. You can't, for example, specify how many worlds there are of a certain sort, even in principle, or precisely when worlds are formed. Everettians argue that, properly understood, this is not a problem, but the detailed clarification is complex and controversial.

There's also a problem in making sense of probability in a multiverse, where everything that can happen does happen. The contemporary way of resolving this problem is to argue that there's a sense in which a rational agent in a world can be uncertain about which branch the agent will end up in after the universal wave function undergoes further branching as the result of a measurement. Such an agent can have rational degrees of belief about the outcomes of quantum measurements, even though all possible outcomes occur on different branches and the agent is split between the branches. Given certain assumptions about what it means to be rational in the face of uncertainty, an agent's degrees of belief will agree with the Born probabilities. The decoherence solution to the basis problem, and the analysis of probability and the derivation of the Born rule, are problematic and the subject of a continuing debate between Everettians and their opponents. See, in particular, the critique by Adrian Kent [36].

5. CONCLUSION

So where does this leave things? In a genuinely indeterministic quantum universe described by a noncommutative mechanics, the quantum probabilities of "what you'll obtain if you measure", which arise as features of the noncommutative structure at the microlevel, must ultimately be related to probabilities of "what's there" in the macroworld in the classical ignorance sense. That's the "small" measurement problem. This works out if the macrosystems used to probe the microworld are modeled as infinite quantum systems, which is how a physical explanation deals with macroscopic descriptions that are insensitive to certain variations at the microlevel.

Critics will object with Bell [37] that this involves FAPP ("for all practical purposes") reasoning, which makes quantum mechanics a predictive instrument rather than a fundamental theory of matter, of interest only to "shopkeepers and engineers", as Einstein put it:

If one wants to consider the quantum theory as final (in principle), then one must believe that a more complete description would be useless because there would be no laws for it. If that were so then physics could only claim the interest of shopkeepers and engineers; the whole thing would be a wretched bungle.

—Albert Einstein ([38], p. 39)

The conclusion here is rather different. The transition from "what you'll obtain if you measure" to "what's there" in a quantum measurement process, where "what's there" is intrinsically random, is like a phase transition. Quantum mechanics*can* model what happens, but certain idealizations in the mathematical description of the macrosystems used to probe the quantum world are unavoidable. This was Bohr's position, that a coherent quantum physics requires that observers and the recording infrastructure should be left out of the quantum theoretical story and described classically:

It is decisive to recognize that, *however far the phenomena transcend the scope of classical physical explanation, the account of all evidence must be expressed in classical terms.* The argument is simply that by the word "experiment" we refer to a situation where we can tell others what we have done and what we have learned and that, therefore, the account of the experimental arrangement and the results of the observations must be expressed in unambiguous language with suitable application of the terminology of classical physics.

—Niels Bohr ([39], p. 209)

Heisenberg added the idea of a movable "cut" between the part of the world doing the measuring and the part being measured. The cut is movable because any part of the world can, in principle, be described purely quantum mechanically, but the idea is that some part of the world—the ultimate measuring and recording device in any application of the theory—must be left out of the quantum mechanical description and described classically.

Bohr's comment is hardly an argument. On the information-theoretic interpretation, the requirement that the experimental arrangement and observational outcomes should be given a "classical" account is understood as the constraint that the ultimate measuring instrument in any quantum experiment should be modeled as a commutative or Boolean system.

To sum up: The "big" measurement problem disappears with the recognition that quantum mechanics is about the structure of information in a universe in which measurement outcomes are intrinsically random events, so there can't be a dynamical account of how a measurement outcome event, indefinite before a measurement, becomes definite. Quantum states, unlike classical states, don't characterize a system in terms of its properties, but rather specify the probabilities of "what you'll obtain if you measure", for example the probabilities that a photon, in a certain state of linear polarization, directed to pass through a beamsplitter, will emerge in the horizontal or vertical channel. The remaining problem is to show how probabilities of "what you'll obtain if you measure" become probabilities of "what's there". This turns out to be a consequence of treating the ultimate measuring instrument in any experimental arrangement designed to probe the quantum world as a commutative or Boolean system. Certain idealizations in the quantum mechanical analysis are unavoidable, but these are analogous to idealizations in the classical treatment of phase transitions.

ACKNOWLEDGMENTS

This article is based on the last Associative Space-Time Sedenions and Their Application in Relativistic Quantum Mechanics and Field Theory

of my book *Bananaworld: Quantum Mechanics for Primates*; Oxford University Press, Oxford, UK, February, 2016.

CONFLICTS OF INTEREST

The author declares no conflict of interest.

REFERENCES

1. Schrödinger, E. Discussion of probability relations between separated systems. *Proc. Camb. Phil. Soci.* **1935**, *31*, 555–563.
2. Colbeck, R.; Renner, R. No extension of quantum theory can have improved predictive power. *Nature Comm.* **2011**, *2*.
3. Colbeck, R.; Renner, R. Is a system's wave function in one-to-one correspondence with its elements of reality? *Phys. Rev. Lett.* **2012**, *108*, 150402.
4. Gleason, A.N. Measures on the closed subspaces of Hilbert space. *J. Math. Mech.* **1957**, *6*, 885–893.

5. Shannon, C. A mathematical theory of communication. *Bell Syst. Tech. J.* **1948**, 379–423, 623–656.

6. Vedral, V. *Decoding Reality*; Oxford University Press: Oxford, UK, 2010.

7. Brown, H.R.; Timpson, C.G. Why special relativity should not be a template for a fundamental reformulation of quantum mechanics. In *Physical Theory and its Interpretation*; Demopoulos, W., Pitowsky, I., Eds.; Springer: New York, NY, USA, 2006; pp. 29–42.

8. Bub, J. Quantum mechanics as a principle theory. *Stud. Hist. Phil. Mod. Phys.* **2000**, *31B*, 75–94.

9. Bub, J. Why the quantum? *Stud. Hist. Phil. Mod. Phys.* **2004**, *35B*, 241–266.

10. Bub, J. Quantum theory is about quantum information. *Found. Phys.* **2005**, *35*, 541–560.

11. Clifton, R.; Bub, J.; Halvorson, H. Characterizing quantum theory in terms of information-theoretic constraints. *Found. Phys.* **2003**, *33*, 1561–1591.

12. Heisenberg, W. Über Quantentheoretischer Umdeutung kinematischer und mechanischer Beziehungen. *Zeit. f. Phys.* **1925**, *33*, 879–893.

13. Cassidy, D.C. *Uncertainty: The Life and Science of Werner Heienberg*; W.H. Freeman: New York, NY, USA, 1992.

14. Born, M.; Jordan, P. Zur Quantenmechanik. *Zeit. f. Phys.* **1925**, *34*, 858–888.

15. Born, M.; Heisenberg, H.; Jordan, P. Zur Quantenmechanik II. *Zeit. f. Phys.* **1926**, *35*, 557–615.

16. Pitowsky, I. Quantum mechanics as a theory of probability. In *Physical Theory and its Interpretation*; Demopoulos, W., Pitowsky, I., Eds.; Springer: New York, NY, USA, 2006; pp. 213–240.

17. Bub, J.; Pitowsky, I. Two dogmas about quantum mechanics. In *Many Worlds? Everett, Quantum Theory and Reality*; Saunders, S., Barrett, J., Kent, A., Wallace, D., Eds.; Oxford University Press: Oxford, UK, 2010.

18. Popescu, S.; Rohrlich, D. Quantum nonlocality as an axiom. *Found. Phys.* **1994**, *24*, 379–385.

19. Pais, A. Einstein and the quantum theory. *Rev. Mod. Phys.* **1979**, *51*, 863–914.

20. Schrödinger, E. Die Gegenwärtige Situation in der Quantenmechanik. *Die Naturwissenschaften* **1935**, *23*, 807–812, 823–828, 844–849.

21. Landsman, N.P. Between classical and quantum. In *Handbook of the Philosophy of Science*; Butterfield, J., Earman, J., Eds.; North-Holland: Amsterdam, The Netherlands, 2007; Volume 2, Philosophy of Physics; pp. 417–553.

22. Landsman, N.P. When champions meet: Rethinking the Bohr-Einstein debate. *Stud. Hist. Phil. Mod. Phys.* **2006**, *37*, 212–241.

23. Carlisle, A.; Kwon, H.; Jeong, H.; Ferraro, A.; Paternostro, M. Limitations of a measurement-assisted optomechanical route to quantum macroscopicity of superposition states. *Phys. Rev. A* **2015**, *92*, 022123.

24. Hepp, K. Quantum theory of measurement and macroscopic observables. *Helv. Phys. Acta* **1972**, *45*, 237–248.

25. Bell, J.S. On wave packet reduction in the Coleman–Hepp model. *Helv. Phys. Acta* **1975**, *48*, 93–98.

26. Bohm, D. A Suggested Interpretation of the Quantum Theory in Terms of "Hidden" Variables, I. *Phys. Rev.* **1952**, *85*, 166–179.

27. Bohm, D. A Suggested Interpretation of the Quantum Theory in Terms of "Hidden" Variables, II. *Phys. Rev.* **1952**, *85*, 180–193.

28. Everett, H. Relative State Formulation of Quantum Mechanics. *Rev. Mod. Phys.* **1957**, *29*, 454–462.

29. Everett, H. The Theory of the Universal Wave Function. In *The Many-Worlds Interpretation of Quantum Mechanics*; De Witt, B., Graham, N., Eds.; Princeton University Press: Princeton, NJ, USA, 1973.

30. Englert, B.-G.; Scully, M.O.; Süsmann, G.; Walther, H. Surrealistic Bohm trajectories. *Zeit. f. Nat. A* **1992**, *47*, 1175–1186.

31. Scully, M.O.; Englert, B.G.; Walther, H. Do Bohm trajectories always provide a trustworthy physical picture of particle motion? *Physica Scripta* **1998**, *76*, 41–46.

32. Dürr, D.; Fusseder, W.; Goldstein, S.; Zanghi, N. Comment on "Surrealistic Bohm Trajectories". *Zeit. f. Nat.* **1993**, *48a*, 1261–1262.

33. Werner, R. Guest post on Bohmian Mechanics. Available online: https://tjoresearchnotes.wordpress.com/2013/05/13/guest-post-on-bohmian-mechanics-by-reinhard-f-werner (accessed on 22 October 2015).

34. Wallace, D. *The Emergent Multiverse: Quantum Theory According to the Everett Interpretation*; Oxford University Press: Oxford, UK, 2012.

35. Wallace, D. *Many Worlds? Everett, Quantum Theory, and Reality*; Saunders, S., Barrett, J., Kent, A., Wallace, D., Eds.; Oxford University Press: Oxford, UK, 2010.

36. Kent, A. One world versus many: The inadequacy of Everettian accounts of evolution, probability, and scientific confirmation. In *Many Worlds? Everett, Quantum Theory, and Reality*; Saunders, S., Barrett, J., Kent, A., Wallace, D., Eds.; Oxford University Press: Oxford, UK, 2010; pp. 307–354.

37. Bell, J.S. Against "measurement". *Phys. World* **1990**, *3*, 33–40.

38. Einstein, A. A letter to Schrödinger dated December 22, 1950. In *Letters on Wave Mechanics*; Przibram, K., Ed.; Philosophical Library: New York, NY, USA, 1967.

39. Bohr, N. Discussions with Einstein on epistemological problems in modern physics. In *Albert Einstein: Philosopher-Scientist*; Schilpp, P.A., Ed.; The Library of Living Philosophers: Evanston, IL, USA, 1949; Volume 7, pp. 201–241.

10
CHAPTER
COMPLEX MATTER SPACE AND RELATIVISTIC QUANTUM MECHANICS

Reza Ahangar

Department of Mathematic, Texas A & M University-Kingsville, Kingsville, TX, USA

ABSTRACT

The Special Relativity Theory cannot recognize speed faster than light. New assumption will be imposed that matter has two intrinsic components, 1) mass, and 2) charge, that is M = m + iq. The mass will be measured by real number system and charged by an imaginary unit. This article presents a Complex Matter Space in Relativistic Quantum Mechanics. We are hoping that this approach will help us to present a general view of energy and momentum in Complex Matter Space. The conclusion of this article on Complex Matter Space (CMS) theory will lead help to a better understanding toward the conversion of mass and energy equation, unifying the forces, and unifying relativity and quantum mechanics.

Keywords: Complex Matter, Complex Mass, Complex Energy and Momentum

1. INTRODUCTION TO THE COMPLEX MATTER SPACE

It is common in mathematics to define an abstract space like metric, Hilbert, or Banach spaces without physical or experimental verifications. In fact logical approach with definitions, some related axioms, and presentation of some examples may be sufficient to accept and apply to practical life problems. But for constructing a theoretical physical space, in addition to the logical and theoretical approach, an experimental or physical justification is needed (see [1] and [2]).

In 1928, the British Physicist Paul Dirac realized that when electrons are moving very close to the speed of light, special relativity needs to be incorporated which lead to particles with negative mass (see [3]). The existence of a particle moving faster than light was presented by many authors including, Sudarshan who discussed the physics of complex mass (see [4]).

During the past decades, various theoretical approaches were presented to unify relativity and quantum mechanics which include complex space-time and axiomatic quantum field theory (see [5] and [6]).

The content of the paper on "Imaginary Relativity" is a right direction toward Complex Matter Space (CMS) but not sufficient for the unifying forces. The CMS postulates may answer some of the questions in this article's concluding remarks. This could be a reason for the statement that "the author does not know where the imaginary world is located or what its nature is?" (see [7]).

We will present postulates that a particle has two components that can be demonstrated in a complex plane as a real component mass (m) and imaginary component charge (q):

$$M = m + iq \tag{1.1}$$

We will show that these assumptions lead us to unify the electromagnetic force and gravitational force.

For example the interaction force between two particles M and M' can be described by

$$F = k\frac{M \cdot M'}{r^2} = k\frac{(m+iq)\cdot(m'+iq')}{r^2} \tag{1.2}$$

We can observe that, after expansion of (1.2), two terms on the right hand side represent gravitational forces and electromagnetic forces. But two other imaginary components seemed to me not to represent any forces.

Assume that the terminology "matter" in general is an undefined term. In this paper, in addition to the postulates of Special Relativity Theory (SRT), we accept the following new postulates:

Complex Matter Space (CMS)	
Postulate (I):	Postulate (II):
Mass and charge are two intrinsic components of the matter and they can be measured and determined in a complex plane. The real part is equal to the mass and the imaginary part is the charge of the particle. That is $M = m + iq$.	In a one dimensional space, position and velocity are assumed to be real numbers and the rate of change of the position of a particle can be assumed to be faster than the speed of light.

The position x(t) and the velocity dx/dt will stay a real number. The speed of light remains a constant real number, c. By this agreement, and contrary to the paper "Imaginary Relativity" by Rezaei, we are not accepting variation of the speed of light or considering it as a complex number. For decades, the discovery of the nature of other forces was a dream. Acceptance of this postulate leads to the necessity to reaffirm the entire physics, Newtonian mechanics, relativity, and quantum mechanics.

The organization of this paper is the following: A new view of complex matter will be presented based on two postulates. We use a geometric interpretation of complex matter to justify new postulates in a complex coordinate plane.

By these new postulates of Mass and Energy we can restructure the relativistic view of complex matter, the Einstein equation of mass-energy. Finally, energy and momentum in Relativistic Quantum Mechanics will be recalculated in the Complex Matter Space.

2. Geometric View of the Complex Matter Using Imaginary Object

Traveling Beyond the Speed of Light: According to the Special Theory of Relativity, we may not be allowed to cross the boundary of the speed of light c in the real world. But if we can pretend that we exist, in the imaginary world then it may be allowed to go faster than light, since our conjugate space will be the real world.

Assume that we are traveling by a rocket which is taking off from the earth to go cruising in space. How much energy is required to accelerate to the speed of light? It would be impossible for a particle to accelerate to a velocity of light c because the speed of light is constant in a vacuum for all inertial observers. What do we observe?

1) Observers in all inertial frames on the earth would measure the same value for c and in fact*infinite energies* are required to accelerate objects to the speed of light.

2) If we can break the wall of speed of light and travel faster than light, an observer would see things going *crazy* around—by SRT time dilation the time would freeze around us—mass will approach infinity.

Experiment of Moving the Image of a Light Faster than the Speed of Light: Assume a source of a light located at the epicenter of two spheres (O, c) and (O, R) where OM = R ? c whose radius R is much greater than the speed of light c. This sphere can be real but radius can be astronomically large. To perform this experiment, one can select c = 1 as a unit of the measurement thus R ? 1.

The source of light at the center (O) emitting the light signal in each direction and rotating with the rate of $\omega(t)=\dfrac{d\theta}{dt}$.

The light is traveling at the speed of the c and it will take seconds to show the image on the dome. We will measure the linear speed of the image on the dome.

For simplicity two circles of radius c = 1 and R with the images of mapping of A and B on the larger circle is M and N, where their arc-lengths are s and s' respectively.

In Figure 1 the light signal emits from the center (O) and transfers the image of the point A to a point M on the second circle, while the point is moving on the circumference of the first circle. In the following simple approach, we will demonstrate that the image point A is moving faster than the point M on the circumference of the second circle.

We denote T(A) = M and similarly T(B) = N such that length of the arc AB is s and T(s) = S.

$$s=r\cdot\theta=c\cdot\theta\Rightarrow\frac{ds}{dt}=\frac{ds}{d\theta}\cdot\frac{d\theta}{dt}=c\frac{d\theta}{dt}$$

Let us assume that the measure of the angle (AOB) = theta varies with the rate of a fixed real number, for example k = 2, therefore $\frac{d\theta}{dt}=2$. Substituting in the previous relation implies that

$$s = r \cdot \theta = c \cdot \theta \Rightarrow \frac{ds}{dt} = c\frac{d\theta}{dt} = 2c$$

Thus, the image M which is moving on the larger circle (O, R) with the linear speed of 2c is faster than light.

Complex Coordinate in Minkowski Space-Time: Absolute time may be an undefined word and we try to postulate, define, and measure it by its changes, variations, and motions with other physical quantities in space, i.e. physical motion of particles, rotation of the earth around the sun, or clock's hands on a circle. The unit of time practically is what we perceive about how much the physical objects change position, velocity, energy, or

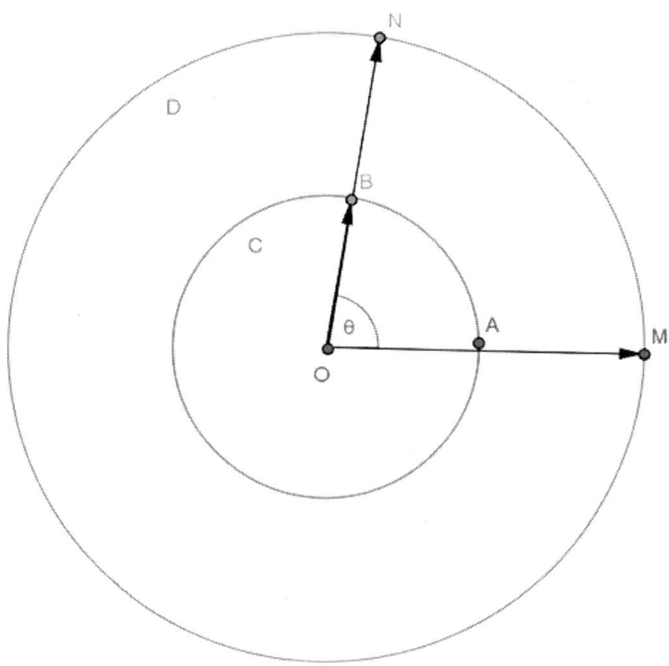

Figure 1. The light signal is emitting from the center (O) and transfers the image of the point A to the second circle M while the point is moving on the circumference of the first circle and the image is moving faster on the circumference of the second circle.

mass. How do we define one hour, one minute, or one second? They are the measurements of the physical changes that we try to demonstrate by arc-length on a clock using the unit of distance or radians on a circle.

Are they real measurements of time? In the following mathematical argument we introduce $i = \sqrt{-1}$ as a unit of time on the imaginary axis. Let us consider a vector $\psi = (u, v, w, t)$ such that

$$\langle \psi_1, \psi_2 \rangle = u_1 u_2 + v_1 v_2 + w_2 w_2 + (it_1)(it_2)$$
$$\langle \psi_1, \psi_2 \rangle = u_1 u_2 + v_1 v_2 + w_2 w_2 - (t_1)(t_2) \tag{2.1}$$

Magnitude of a Vector 4D-Complex Matter Space: We may use the norm in the inner product space as a distance or magnitude of a vector $\psi = (u, v, w, t)$

$$\langle \psi, \psi \rangle = \|\psi\|^2 = u^2 + v^2 + w^2 - t^2$$

Verifying the CMS Postulate (II): According to time and length dilation in SRT, the value of time and length variation

$$\Delta t = \frac{\Delta t'}{\sqrt{1 - \left(\frac{v}{c}\right)^2}} = \gamma \cdot \Delta t' \text{ and } \Delta L = \frac{\Delta L'}{\sqrt{1 - \left(\frac{v}{c}\right)^2}} = \gamma \cdot \Delta L' \tag{2.2}$$

such that

(a) $\gamma = \frac{1}{\sqrt{1 - \left(\frac{v}{c}\right)^2}} = \frac{1}{i\sqrt{\left(\frac{v}{c}\right)^2 - 1}} = \frac{-i}{\sqrt{\left(\frac{v}{c}\right)^2 - 1}}$ and (b) $\gamma^* = \frac{i}{\sqrt{\left(\frac{v}{c}\right)^2 - 1}}$ (2.3)

can be interpreted for the velocity v as a real number greater than the speed of light.

This is one indication that time may have an imaginary unit.

Assuming a complex time variation

$t_1 = a + ib$ and

The relation (2.2) can be expressed by

$$\Delta t = \frac{p+iq}{\sqrt{1-\left(\frac{v}{c}\right)^2}} = \gamma \cdot (p+iq)$$

Case (I): the motion velocity v < c.

The values of α , β , and γ will remain real numbers. Consequently, according to (2.2) velocity v and distance L will be real numbers.

$$v(t) = \lim_{\Delta t \to 0} \frac{\Delta L}{\Delta t} = \lim_{\Delta t \to 0} \frac{\gamma \cdot \Delta L'}{\gamma \cdot \Delta t'} = \lim_{\Delta t \to 0} \frac{\Delta L'}{\Delta t'}$$

Case (II): the motion velocity v > c: If the particle velocity is faster than the speed of light (v > c) then both γ and γ^* are pure imaginary numbers and using (2.2) the following shows that

$$v(t) = \lim_{\Delta t \to 0} \frac{\Delta L}{\Delta t} = \lim_{\Delta t \to 0} \frac{\gamma \cdot \Delta L'}{\gamma \cdot \Delta t'} = \lim_{\Delta t \to 0} \frac{\Delta L'}{\Delta t'}$$

is a real number. The argument presented in both cases based on the inequality v > c or v < c, demonstrates that the relation in an ordered pairs of real numbers. Thus we cannot select these two numbers in the complex plane.

3. BEYOND THE SPEED LIMIT OF LIGHT AND RELATIVISTIC QUANTUM VIEW OF THE COMPLEX MATTER SPACE

Having a new unified presentation of mass and charge in the form of complex matter will bring a new question for us. How do we interpret the principles of Relativistic Quantum Mechanics?

Antimatter: Substance composed of elementary particles having the mass and electric charge of ordinary matter (such as electrons and protons) but for which the charge and related magnetic properties are opposite in sign. The existence of antimatter was predicted by the electron theory of P. A. M. Dirac. In 1932 the positron (antielectron) was detected in cosmic rays, followed by the antiproton and the antineutron detected through the use of particle accelerators. Positrons, antiprotons, and antineutrons, collectively called antiparticles, are the antiparticles of electrons, protons, and neutrons, respectively. When matter and antimatter are in close proximity, annihilation occurs within a fraction of a second, releasing large amounts of energy. The (relativistic) mass of an object measured by an observer in the xyz-frame is given by

$$M(t) = \frac{M_{\text{rest}}}{\sqrt{1-\left(\frac{v}{c}\right)^2}} = \frac{m_0 + iq_0}{\sqrt{1-\left(\frac{v}{c}\right)^2}} \tag{3.1}$$

By accepting the complex matter Postulate (I), the velocity in (3.1) can be any real or complex number. Mathematically v could be greater than the speed of light, but the mass m would become imaginary. Physically we would have to get to the speed of light first i.e. v = c, which gives us an undefined value for M. So we believe that nothing moves faster than the speed of light because we do not like observables to be imaginary.

A hypothetical particle that is traveling faster-than a light particle is consistent with the special theory of relativity. According to this theory, a free particle has an energy E and a momentum p which form a Lorentz four-vector. The length of this vector is a scalar, having the same value in all inertial reference frames. One writes

$$E^2 = c^2P^2 + m^2c^4 \tag{3.2}$$

Equation (3.2), where c is the speed of light and the parameter m^2 is a property of the particle, independent of its momentum and energy. Three cases may be considered: m^2 may be positive, zero, or negative. The case $m^2 > 0$ applies for atoms, nuclei, and the macroscopic objects of everyday experience. The positive root m is called the rest-mass. If $m^2 = 0$, the particle is called massless. A few of these are known: the electron neutrino, the muon neutrino, the photon, and the graviton. The name tachyons (after a Greek word for swift) has been given to particles with $m^2 < 0$.

Mayer in 2007, designed the energy momentum in a complex plane very well without acknowledging the idea of complex matter considering the speed of light c = 1 a natural unit (see [8]). The relation (3.2) can be interpreted by.

Energy in the complex plane: A vector form of the energy is demonstrated with the horizontal component OH = mc^2, and vertical component ZH = pc. The state of the energy and its magnitude can be described by

$$\boldsymbol{E} = \boldsymbol{m_0c^2} + \boldsymbol{ipc} \tag{3.3}$$

$$\|E\| = \sqrt{\left(m_0 c^2\right)^2 + (pc)^2} \tag{3.4}$$

According to postulate (I), Figure 2 and Figure 3 demonstrate that the energy will have the two components of real and imaginary values. The energy for mass-less particles can be observed on the diagram when $\|E\| = pc$. When the real part of particle mass is not zero we will derive the energy, mass, and charge in relativistic mechanics. The norm (modulus) of the particle-like with speed faster than light can be described by

$$\|M\| = \frac{\|M_0\|}{\left\|\sqrt{1-\left(\frac{v}{c}\right)^2}\right\|} = \frac{\sqrt{m_0^2+q_0^2}}{\sqrt{1-\left(\frac{v}{c}\right)^2}} = \frac{\sqrt{m_0^2+q_0^2}}{i\sqrt{\left(\frac{v}{c}\right)^2-1}} = \gamma\sqrt{m_0^2+q_0^2} = -i\bar{\gamma}\sqrt{m_0^2+q_0^2}$$

where, the magnitude of the complex vector for particle and factor gamma is defined by:

$$\|M\| = \gamma\sqrt{m_0^2+q_0^2} \quad \text{where } \gamma = \frac{1}{\sqrt{1-\left(\frac{v}{c}\right)^2}} \tag{3.5}$$

The following steps lead to the effects of complex matter theory on elementary quantum mechanics (see [9] and [10]).

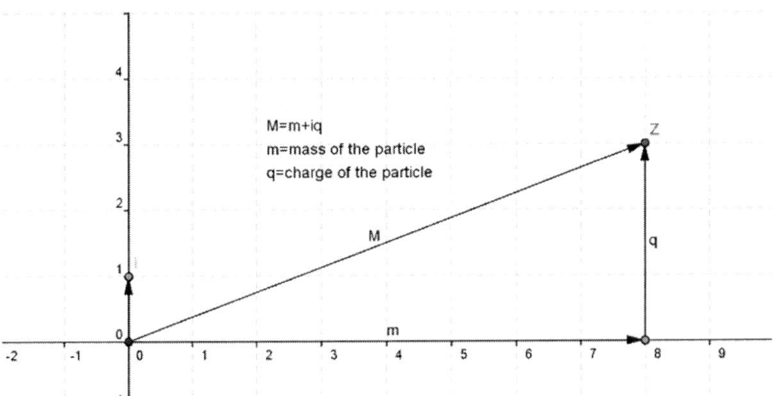

Figure 2. The state of the complex matter in a complex plane with two components (m, q) and it is denoted by $M = m + iq = re^{i\theta}$ where $r = \sqrt{m^2+q^2}$, and $\theta = Arg\left(\frac{q}{m}\right)$.

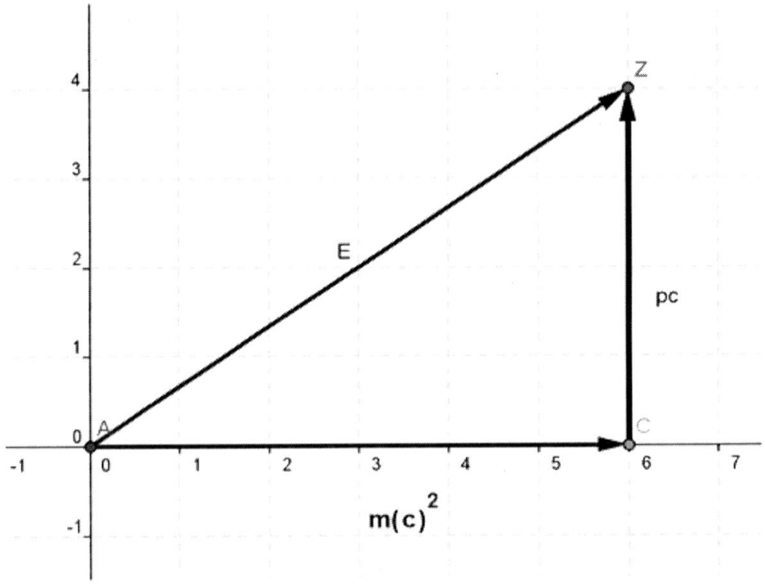

Figure 3. Energy in the complex plane: A vector form of the energy demonstrated with the horizontal component OH = mc^2, and vertical component ZH = pc. Energy vector $\boldsymbol{E = mc^2 + ipc}$ with a magnitude $\|E\| = \sqrt{(m_0c^2)^2 + (pc)^2}$.

It is well known that the energy of a photon can be calculated if we assume that light is composed of particles.

The energy is proportional to the density of the photons particles, that is

1) Photon is a complex particle with zero mass. Energy of a package of photon particles is equal to energy of each particle multiply by the number of particles (frequency), that is

$$E = h \cdot f_q \tag{3.6}$$

Where, h is the Plank constant real number. The momentum can be calculated using.

2) Campton Effect when particle—like is a wave. Momentum of the of the photon wave will be equal to Energy/speed, that is

$$p_{real} = \frac{E}{c} = \frac{h \cdot f_q}{c} \tag{3.7}$$

By considering that velocity is equal to frequency multiplied by the wave length, and $c = \lambda \cdot f_q$ we will describe the momentum propagation of the photons as a wave by

$$p_{imag} = \frac{h \cdot f_q}{c} = \frac{h \cdot f_q}{\lambda \cdot f_q} = \frac{h}{\lambda} \tag{3.8}$$

3) Using de Broglie's Law we can find the energy of the wave (by 3.6) and replacing the frequency f_q is equal to the energy of the mass, that is $E_{imag} = h \cdot f_q = mc^2$.

Wave length $\lambda = \frac{h}{p} = \frac{h}{mv}$. Combining $\lambda = \frac{h}{p}$ and $E = h \cdot f = mc^2$ implies that: $w = \lambda f = \frac{h}{p} \cdot \frac{pc}{h} = c$. The wave frequency can be interpreted by $\frac{1}{\lambda} = f = \frac{p}{h}$. This relation can also be extended to the Einstein mass- energy equivalence

$$\lambda = \frac{h}{p} = \frac{hc^2}{mvc^2} = \frac{hc^2}{E \cdot v} \Rightarrow \frac{p}{E} = \frac{v}{c^2} \text{ or } \frac{E}{p} = \frac{c^2}{v} \tag{3.9}$$

Consequently:

$$\frac{pc}{E} = \frac{v}{c} \tag{3.10}$$

Let E = mv²/2 and it is clear that $dE/dt = mv = p$. We use (7.4) to modify our definition of the angle alpha in (Figure 3): $\alpha = \sin^{-1}\left(\frac{pc}{E}\right) = \sin^{-1}\left(\frac{v}{c}\right)$.

4. MODIFYING THE EINSTEIN'S EQUATION IN COMPLEX MATTER SPACE

Using Postulate (I) let us describe a particle of a Complex Matter Space by a set of points $M(t) = m(t) + iq(t)$ where m is the mass and q is the charge at time t in the interval (a, b) for local real parameters a and b.

We will describe the instantaneous rates of changes of this complex matter by $\frac{dM}{dt} = \frac{dm}{dt} + i\frac{dq}{dt}$.

Complex Matter-Instantaneous Rate of Changes: Given a complex matter in the Equation (3.1) then the instantaneous rate of change satisfies the following relation

$$\frac{dM}{dt} = \frac{vM}{(c^2 - v^2)} \frac{dv}{dt} \text{ or } dv = \frac{(c^2 - v^2)}{vM} dM \tag{4.1}$$

Proof: Assuming $M = m + iq$ and taking the derivative of (3.1)

$\frac{dM}{dv} = \frac{vM_0}{c^2}\left(\frac{c^2}{c^2 - v^2}\right)^{3/2} = \frac{vM_0}{c^2 - v^2}\sqrt{\frac{c^2}{c^2 - v^2}}$ implies

This will conclude the relation (4.1) and also $dv = \frac{(c^2 - v^2)}{vM} dM$.

Energy Equation in Complex Matter Space: The instantaneous rate of change of Energy in Complex Matter Space can be described by

$$dE = c^2 dm + ic^2 dq \tag{4.2}$$

Justification: Let us assume, in the following approach, that the definition of energy and force for the Complex Matter Space is staying the same. Thus;

$$dE = F dx = \frac{dp}{dt} dx = dp \frac{dx}{dt} = (dp)v \tag{4.3}$$

where, the velocity $v = \frac{dx}{dt}$ is a real vector function, but the momentum $p = M \cdot v$ will be a complex vector function. The differentiation of momentum will be

$$\mathrm{d}p = M\mathrm{d}v + v\mathrm{d}M \tag{4.4}$$

Substitute (4.4) in the relation (4.3)

$$\mathrm{d}E = \left[M\mathrm{d}v + v\mathrm{d}M\right]v = v^2\mathrm{d}M + vM\frac{\left(c^2 - v^2\right)\mathrm{d}M}{vM}$$

Simplifying the right hand side concludes

$$\mathrm{d}E = c^2\mathrm{d}M = c^2\mathrm{d}m + ic^2\mathrm{d}q \tag{4.5}$$

To explain the relativistic energy in Complex Matter Space, we use the trigonometric values in the complex plane (using the relation (3.10) and Figure 3);

$\alpha = \sin^{-1}\left(\dfrac{pc}{E}\right) = \sin^{-1}\left(\dfrac{v}{c}\right)$ and

assume $\beta = \dfrac{v}{c}$ then $\cos(\alpha) = \sqrt{1-\sin^2(\alpha)} = \sqrt{1-\beta^2}$

When the real part of particle mass is not zero we will derive the energy, mass, and charge in relativistic mechanics.

The complex point E can be described in the following form

$$\boldsymbol{E} = \boldsymbol{mc}^2 + \boldsymbol{ipc} = \left\|E\right\|\mathrm{e}^{i\alpha} = \sqrt{\left(m_0c^2\right)^2 + (pc)^2}\left(\cos(\alpha) + i\sin(\alpha)\right)$$

Replacing cosine and sine of angle alpha from the previous step, we will get an energy state description

$$\boldsymbol{E} = \sqrt{\left(m_0c^2\right)^2 + (pc)^2}\left(\sqrt{1-\left(\frac{v}{c}\right)^2} + i\frac{v}{c}\right) \tag{4.6}$$

This is a correction of the conclusion presented in a paper by Mayer in 2007 (see [11]). The relation (4.6) is the state of the energy of a particle with the real rest mass moving faster than light.

5. STATE OF THE MOMENTUM IN COMPLEX MATTER SPACE

If the charge in wave-like particle is zero, then dividing the real part of energy by real part of momentum will produce the following relations

$$p_{real} = mv = \frac{m_0 v}{\sqrt{1-\left(\frac{v}{c}\right)^2}} \text{ and } E_{real} = mc^2 = \frac{m_0 c^2}{\sqrt{1-\left(\frac{v}{c}\right)^2}} \tag{5.1}$$

and we obtain $\frac{p_{real}}{E_{real}} = \frac{v}{c^2}$ which implies the relative velocities $\frac{v}{c} = \left\|\frac{pc}{E}\right\|$.

Using Figure 2 and the relation (3.10)

$$\frac{v}{c} = \frac{pc}{E} = \sin(\alpha) = \beta \tag{5.2}$$

The State of the Complex Matter: Using relations (5.2), one can find the value of gamma from (2.2)

$$\|M\| = \gamma\sqrt{m_0^2 + q_0^2} = \sqrt{m_0^2 + q_0^2}\,\sec(\alpha) \text{ where } \alpha = \sin^{-1}\left(\frac{v}{c}\right) \tag{5.3}$$

State of the Complex Energy: The complex point E can be described in the following state of energy in the complex model $\boldsymbol{E} = \boldsymbol{mc}^2 + \boldsymbol{ipc} = \|\boldsymbol{E}\|\mathrm{e}^{i\alpha}$ Þ

$$\boldsymbol{E} = \sqrt{\left(m_0 c^2\right)^2 + (pc)^2}\left(\cos(\alpha) + i\sin(\alpha)\right) \tag{5.4}$$

Thus, from the relation (5.4) one can conclude the real and imaginary components of energy to be

$$E_{real} = \sqrt{\left(m_0 c^2\right)^2 + (pc)^2}\cos(\alpha), \text{ and } E_{imag} = \sqrt{\left(m_0 c^2\right)^2 + (pc)^2}\sin(\alpha) \tag{5.5}$$

(see [12]).

The State of the Momentum of Complex Matter: Assume that $M_0 = (m_0, q_0)$ represents the rest point at the initial time with two components. According to postulate (II), a moving particle in the complex plane at the speed $v = \|V(t)\| = \left\|\frac{dx}{dt}\right\|$ on the one dimensional x-axis, will be a real function of time. The momentum at time t

can be described by

$$\|p\| = rv = \|M\|v = \frac{\|M_0\|v}{\sqrt{1-\frac{v^2}{c^2}}} = \frac{\sqrt{m_0^2+q_0^2}\,v}{\sqrt{1-\frac{v^2}{c^2}}} = v\sqrt{m_0^2+q_0^2}\sec(\alpha) \tag{5.6}$$

where, $\frac{1}{\sqrt{1-\left(\frac{v}{c}\right)^2}} = \frac{1}{\sqrt{1-\beta^2}} = \frac{1}{\cos(\alpha)} = \sec(\alpha)$. As a result the wave length can be described

$$\lambda = \frac{h}{\|p\|} = \frac{h}{v\sqrt{m_0^2+q_0^2}\sec(\alpha)} = \frac{h\cdot\cos(\alpha)}{v\sqrt{m_0^2+q_0^2}} \tag{5.7}$$

6. DISCUSSIONS AND CONCLUSION

According to the original postulate, we determine a point in a complex plane by $\boldsymbol{M} = \|\boldsymbol{M}\|e^{i\theta}$ where,

$\theta = Arg\left(\frac{q}{m}\right)$. One can explore and study the implication of this assumption, and demonstrate the consistency

with other phenomenon, theories, or experiments.

The existence of a particle traveling faster than light has been presented by many authors including, Sudarshan who discussed the Physics of complex mass (see [12]).

The idea of negative matter appears in past theories of matter that have now been abandoned (see [13]). Using the once popular vortex theory of gravity, the possibility of matter with negative gravity was discussed by William Hicks in the 1880s.

In 1928, British physicist Paul A. M. Dirac generalized Einstein's famous equation $E = mc^2$ by considering negative mass. This hypothesis allowed for the existence of anti-particles in our universe.

These anti-particles are, literally, mirror images of normal or real matter. Each anti-particle has the same mass as its corresponding particle, but the electrical charges are reversed. Examples of antimatter discoveries of the 20th century are: 1) Positrons—Electrons with a positive instead of negative charge; 2) Anti-protons—Protons that have a negative instead of the usual positive charge; 3) Anti-atoms—Pairing together positrons and antiprotons (see [14]).

Scientists at the European Organization for Nuclear Research, created the first anti-atom. Nine anti-hydrogen atoms were created, each lasting only 40 nanoseconds. As of 1998, CERN researchers were pushing the production of anti-hydrogen atoms to 2000 per hour.

Both particles involved in the interaction process are completely annihilated and create a very huge explosion that occurs when antimatter and matter interact transferring the entire mass of both objects into energy.

It is part of the dream unification theory to build a matter-antimatter reaction engine?

7. FUTURE RESEARCH

The quantum states of the particles can be studied in the Complex Matter Space through interactions with other fundamental particles.

The Electromagnetic Interaction, Weak Interaction, and Strong Interaction forces can be considered in the Complex Matter Space. The standard model in quantum state does not consider Gravitation as part of the standard model.

The Complex Matter Space model can include all of the interactions as part of the model including the Weak Forces.

REFERENCES

1. Naber George, L. (1992) The Geometry of Minkowski Spacetime. An Introduction to the Mathematics of Special Theory of Relativity. Dover.
2. David, H. and Rensick (with the Assistance of Farrell Edwards and John Merrill) (1970) Fundamentals of Physics. John Wiley & Sons, Inc., Hoboken.

3. Dirac, P.A.M. (1928) The Quantum Theory of Electron. Proceedings of the Royal Society A, 117, 610-624.
4. Sudarshan, E.C.G. (1970) Physics of Complex Mass Particles. University of Texas, Austin.
5. Gerald, K. (2003) Quantum Physics, Relativity, and Complex Space Time: Toward a New Synthesis. (Originally published in 1990 by North-Holland, Amsterdam)
6. Richard, F. (1963) The Feynman Lectures on Physics. Vol. III, Addison Wesley, Reading, 1-6.
7. Mehran, R. (2010) Imaginary Relativity. Wseas Transaction on Communications, 9.
8. Mayer Alexander, F. (2007) Wave Energy in Quantum Mechanics. Journal of Physics, Conference Series, 70, Article ID: 012013. http://iopscience.iop.org/1712-6596/70/1/02013.
9. Wall Ernst, L. (1995) The Vortex Theory of the Electron and its de Broglie Waves Pulse. Institute for Basic Research, Palm Harbor. www.gwphysics.com www.tachyonmodel.com
10. John, W. and Wojceich, Z. (1983) The Quantum Postulate and the Recent Development of Atomic Theory. In: Quantum Theory and Measurement, Princeton University Press, Princeton, 111.
11. Reza, A. (2014) Quantum Complex Matter Space. International Journal of Theoretical and Mathematical Physics, 4, 159-163.
12. Zhang, T.X. (2008) Electric Charge as a form of Imaginary Energy. Progress in Physics, 2.
13. Sudarshan, E.C.G. (1968) A New Formulation of Relativistic Quantum Theory of Fields with Applications to Particles Traveling Faster Than Light. Proceedings of Nobel Symposium on Elementary Particle Theory.
14. Taylor, T.J.R., Zafirots, C.D. and Dubson, M.A. (2004) Modern Physics for Scientists and Engineers. 2nd Edition, Pearson-Printice Hall.

11
CHAPTER

ASSOCIATIVE SPACE-TIME SEDENIONS AND THEIR APPLICATION IN RELATIVISTIC QUANTUM MECHANICS AND FIELD THEORY

Victor L. Mironov[1,2], Sergey V. Mironov[1]

[1]*Institute for Physics of Microstructures RAS, Nizhny Novgorod, Russia*

[2]*Lobachevsky State University of Nizhny Novgorod, Nizhny Novgorod, Russia*

ABSTRACT

We present an alternative sixteen-component hypercomplex scalar-vector values named "space- time sedenions", generating associative noncommutative space-time Clifford algebra. The generalization of relativistic quantum mechanics and field theory equations based on sedenionic wave function and space-time operators is discussed.

Keywords: Clifford Algebra, Space-Time Sedenions, Relativistic Quantum Mechanics, Sedenionic Klein-Gordon Equation, Sedenionic Dirac Equation, Sedenionic Maxwell Equiations

1. INTRODUCTION

The multicomponent hypercomplex numbers such as quaternions and octonions are widely used for the reformulation of quantum mechanics and field theory equations. The first generalization of quantum mechanics and electrodynamics was made on the basis of four-component quaternions, which were interpreted as scalar-vector structures [1] -[5] . The next step was taken on the basis of eight-component octonions, which were interpreted as the sum of scalar, pseudoscalar, polar vector and axial vector [6] - [11] . Scalars and axial vectors are not transformed under spatial inversion, while pseudoscalars and polar vectors change their sign under spatial inversion. Therefore, this interpretation takes only the symmetry with respect to the spatial inversion into account. However, a consistent relativistic approach requires taking full time and space symmetries into consideration that leads to the sixteen-component space-time algebras.

The well-known sixteen-component hypercomplex numbers, sedenions, are obtained from octonions by the Cayley-Dickson extension procedure [12] [13] . In this case the sedenion is defined as

$$S = O_1 + O_2 \boldsymbol{e} \quad , (1)$$

where O_i is an octonion and the parameter of duplication e is similar to imaginary unit $\boldsymbol{e}^2 = -1$. The algebra of sedenions has the specific rules of multiplication. The product of two sedenions

$$S_1 = O_{11} + O_{12}\boldsymbol{e} ,$$

$$S_2 = O_{21} + O_{22}\boldsymbol{e} ,$$

is defined as

$$S_1 S_2 = (O_{11} + O_{12}\boldsymbol{e})(O_{21} + O_{22}\boldsymbol{e}) = (O_{11}O_{21} - \bar{O}_{22}O_{12}) + (O_{22}O_{11} + O_{12}\bar{O}_{21})\boldsymbol{e} , \quad (2)$$

where $\hat{q} = q_0\mathbf{a_0} + q_1\mathbf{a_1} + q_2\mathbf{a_2} + q_3\mathbf{a_3}$ is conjugated octonion. The sedenionic multiplication (2) allows one to introduce a well-defined norm of sedenion. However, such procedure of constructing the higher hypercomplex numbers leads to the fact that the sedenions as well as octonions generate normed but nonassociative algebra [14] - [16] . It complicates the use of the Cayley-Dickson sedenions in the physical applications.

Recently we have developed an alternative approach to constructing the multicomponent values based on our scalar-vector conception realized in associative eight-component octons [17] - [19] and sixteen-component sedeons [20] - [24] . In particular, we have demonstrated the method, which allows one to reformulate the equations of relativistic quantum mechanics and field theory on the basis of sedeonic space-time operators and scalar-vector wave functions. In this paper we present an alternative version of the sixteen-component associative space-time hypercomplex algebra and demonstrate some of its application to the generalization of relativistic quantum mechanics and field theory equations.

2. SEDENIONIC SPACE-TIME ALGEBRA

It is known, the quaternion is a four-component object

$$q_{\nu},\tag{3}$$

where components $\nu = 0,1,2,3$ (Greek indexes $a_0 \equiv 1$) are numbers (complex in general), $\mathbf{e}_0 \equiv 1$ is scalar units and values a_m (Latin indexes m = 1, 2, 3) are quaternionic units, which are interpreted as unit vectors. The rules of multiplication and commutation for a_m are presented in Table 1. We introduce also the space-time basis e_t, e_r, e_{tr}, which is responsible for the space-time inversions. The indexes t and r indicate the transformations (t for time inversion and r for spatial inversion), which change the corresponding values. The value $\mathbf{e}_1 \equiv \mathbf{e}_t$ is a scalar unit. For convenience we introduce numerical designations $\mathbf{e}_2 \equiv \mathbf{e}_r$ (time scalar unit);

$$\mathbf{a}_n\mathbf{a}_m = -\mathbf{a}_m\mathbf{a}_n,$$

$\mathbf{e}_3 \equiv \mathbf{e}_{tr}$ (space scalar unit) and $\mathbf{e}_n\mathbf{e}_m = -\mathbf{e}_m\mathbf{e}_n,$ (space-time scalar unit). The rules of multiplication and commutation for this basis we choose similar to the rules for quaternionic units (see Table 2).

Table 1. Multiplication rules for unit vectors $\mathbf{a}_m$.

	$\mathbf{a}_1$	$\mathbf{a}_2$	$\mathbf{a}_3$
$\mathbf{a}_1$	-1	$\mathbf{a}_3$	$-\mathbf{a}_2$
$\mathbf{a}_2$	$-\mathbf{a}_2$	-1	$\mathbf{a}_1$
$\mathbf{a}_3$	$\mathbf{a}_2$	$-\mathbf{a}_1$	-1

Table 2. Multiplication rules for space-time units.

	$\mathbf{e}_1$	$\mathbf{e}_2$	$\mathbf{e}_3$
$\mathbf{e}_1$	-1	$\mathbf{e}_3$	$-\mathbf{e}_2$
$\mathbf{e}_2$	$-\mathbf{e}_2$	-1	$\mathbf{e}_1$
$\mathbf{e}_3$	$\mathbf{e}_2$	$-\mathbf{e}_1$	-1

Note that the unit vectors $\mathbf{a}_1$, $\mathbf{a}_2$, $\mathbf{a}_3$ and the space-time units $\mathbf{e}_1$, $\mathbf{e}_2$, $\mathbf{e}_3$ generate the anticommutative algebras:

$$\mathbf{n} \neq \mathbf{m} \tag{4}$$

for $\mathbf{e}_\mathbf{n}\mathbf{a}_\mathbf{m} = \mathbf{a}_\mathbf{m}\mathbf{e}_\mathbf{n}$, but $\mathbf{e}_1$, $\mathbf{e}_2$, $\mathbf{e}_3$ commute with $\mathbf{a}_1$, $\mathbf{a}_2$, $\mathbf{a}_3$:

$$\begin{aligned}\tilde{V} = {} & \mathbf{e}_0\left(V_{00}\mathbf{a}_0 + V_{01}\mathbf{a}_1 + V_{02}\mathbf{a}_2 + V_{03}\mathbf{a}_3\right) + \mathbf{e}_1\left(V_{10}\mathbf{a}_0 + V_{11}\mathbf{a}_1 + V_{12}\mathbf{a}_2 + V_{13}\mathbf{a}_3\right) \\ & + \mathbf{e}_2\left(V_{20}\mathbf{a}_0 + V_{21}\mathbf{a}_1 + V_{22}\mathbf{a}_2 + V_{23}\mathbf{a}_3\right) + \mathbf{e}_3\left(V_{30}\mathbf{a}_0 + V_{31}\mathbf{a}_1 + V_{32}\mathbf{a}_2 + V_{33}\mathbf{a}_3\right),\end{aligned} \tag{5}$$

for any n and m. Besides, we assume the associativity of $\mathbf{e}_1$, $\mathbf{e}_2$, $\mathbf{e}_3$, $\mathbf{a}_1$, $\mathbf{a}_2$, $\mathbf{a}_3$ multiplication.

Then we can introduce the sixteen-component space-time sedenion $V_{\nu\mu}$ in the following form:

$$
\begin{aligned}
&V = \mathbf{e}_0 V_{00} \mathbf{a}_0,\\
&\mathbf{V} = \mathbf{e}_0 \left(V_{01}\mathbf{a}_1 + V_{02}\mathbf{a}_2 + V_{03}\mathbf{a}_3 \right),\\
&V_{\mathbf{t}} \equiv V_{\mathbf{1}} = \mathbf{e}_1 V_{10} \mathbf{a}_0,\\
&\mathbf{V}_{\mathbf{t}} \equiv \mathbf{V}_{\mathbf{1}} = \mathbf{e}_1 \left(V_{11}\mathbf{a}_1 + V_{12}\mathbf{a}_2 + V_{13}\mathbf{a}_3 \right),\\
&V_{\mathbf{r}} \equiv V_{\mathbf{2}} = \mathbf{e}_2 V_{20} \mathbf{a}_0,\\
&\mathbf{V}_{\mathbf{r}} \equiv \mathbf{V}_{\mathbf{2}} = \mathbf{e}_2 \left(V_{21}\mathbf{a}_1 + V_{22}\mathbf{a}_2 + V_{23}\mathbf{a}_3 \right),\\
&V_{\mathbf{tr}} \equiv V_{\mathbf{3}} = \mathbf{e}_3 V_{30} \mathbf{a}_0,\\
&\mathbf{V}_{\mathbf{tr}} \equiv \mathbf{V}_{\mathbf{3}} = \mathbf{e}_3 \left(V_{31}\mathbf{a}_1 + V_{32}\mathbf{a}_2 + V_{33}\mathbf{a}_3 \right).
\end{aligned}
\tag{6}
$$

The sedenionic components $\tilde{V} = V + \mathbf{V} + V_{\mathbf{t}} + \mathbf{V}_{\mathbf{t}} + V_{\mathbf{r}} + \mathbf{V}_{\mathbf{r}} + V_{\mathbf{tr}} + \mathbf{V}_{\mathbf{tr}}$ are numbers (complex in general). Introducing designation of scalar and vector values in accordance with the following relations

$$
(V) \tag{7}
$$

we can represent the sedenion in the following scalar-vector form:

$$
(\mathbf{V}) . \tag{8}
$$

Thus, the sedenionic algebra encloses four groups of values, which are differed with respect to spatial and time inversion.

1) Absolute scalars $(V_{\mathbf{t}})$ and absolute vectors $(\mathbf{V}_{\mathbf{t}})$ are not transformed under spatial and time inversion.

2) Time scalars $(V_{\mathbf{r}})$ and time vectors $(\mathbf{V}_{\mathbf{r}})$ are changed (in sign) under time inversion and are not transformed under spatial inversion.

3) Space scalars $(V_{\mathbf{tr}})$ and space vectors $(\mathbf{V}_{\mathbf{tr}})$ are changed under spatial inversion and are not transformed under time inversion.

$$\bar{V}_0 = V_{00} + V_{01}\mathbf{a}_1 + V_{02}\mathbf{a}_2 + V_{03}\mathbf{a}_3,$$
$$\bar{V}_1 = V_{10} + V_{11}\mathbf{a}_1 + V_{12}\mathbf{a}_2 + V_{13}\mathbf{a}_3,$$
$$\bar{V}_2 = V_{20} + V_{21}\mathbf{a}_1 + V_{22}\mathbf{a}_2 + V_{23}\mathbf{a}_3,$$

4) Space-time scalars $\bar{V}_3 = V_{30} + V_{31}\mathbf{a}_1 + V_{32}\mathbf{a}_2 + V_{33}\mathbf{a}_3.$ and space-time vectors $\tilde{V} = \bar{V}_0 + \mathbf{e}_1\bar{V}_1 + \mathbf{e}_2\bar{V}_2 + \mathbf{e}_3\bar{V}_3$ are changed under spatial and time inversion.

Further we will use the symbol 1 instead units a_0 and e_0 for simplicity. Introducing the designations of scalar- vector values

$$\begin{aligned} V_0 &= (V_{00} + \mathbf{e}_1 V_{10} + \mathbf{e}_2 V_{20} + \mathbf{e}_3 V_{30}), \\ V_1 &= (V_{01} + \mathbf{e}_1 V_{11} + \mathbf{e}_2 V_{21} + \mathbf{e}_3 V_{31}), \\ V_2 &= (V_{02} + \mathbf{e}_1 V_{12} + \mathbf{e}_2 V_{22} + \mathbf{e}_3 V_{32}), \\ V_3 &= (V_{03} + \mathbf{e}_1 V_{13} + \mathbf{e}_2 V_{23} + \mathbf{e}_3 V_{33}). \end{aligned} \quad (9)$$

we can write the sedenion (6) in the following compact form:

$$\tilde{V} = V_0 + V_1\mathbf{a}_1 + V_2\mathbf{a}_2 + V_3\mathbf{a}_3 \quad . (10)$$

On the other hand, introducing designations of space-time sedenion-scalars

$$V = V + V_t + V_r + V_{tr} = V_1\mathbf{a}_1 + V_2\mathbf{a}_2 + V_3\mathbf{a}_3 \quad (11)$$

we can write the sedenion (6) as

$$\tilde{V} = V_0 + V \quad , (12)$$

or introducing the sedenion-vector

$$\tilde{A}\tilde{B} = (A_0 + A)(B_0 + B) = A_0 B_0 + A_0 B + A B_0 + (A \cdot B) + [A \times B] \quad , (13)$$

we can rewrite the sedenion in following compact form:

$$(A \cdot B) = -A_1 B_1 - A_2 B_2 - A_3 B_3 \quad . (14)$$

Further we will indicate sedenion-scalars and sedenion-vectors with the bold capital letters.

Let us consider the sedenionic multiplication in detail. The sedenionic product of two sedenions $[A\times B]=(A_2B_3-A_3B_2)\mathbf{a}_1+(A_3B_1-A_1B_3)\mathbf{a}_2+(A_1B_2-A_2B_1)\mathbf{a}_3$ and $\tilde{F}=\tilde{A}\tilde{B}=F_0+F$ can be represented in the following form

$$\begin{aligned}F_0 &= A_0B_0 - A_1B_1 - A_2B_2 - A_3B_3,\\ F_1 &= A_1B_0 + A_0B_1 + (A_2B_3 - A_3B_2),\\ F_2 &= A_2B_0 + A_0B_2 + (A_3B_1 - A_1B_3),\\ F_3 &= A_3B_0 + A_0B_3 + (A_1B_2 - A_2B_1).\end{aligned} \quad . (15)$$

Here we denoted the sedenionic scalar multiplication of two sedenion-vectors (internal product) by symbol "·" and round brackets

$$A^2=(A\cdot A)=-A_1^2-A_2^2-A_3^2 \quad , (16)$$

and sedenionic vector multiplication (external product) by symbol "×" and square brackets,

$$|A|^2=-(A\cdot A)=A_1^2+A_2^2+A_3^2 \quad . (17)$$

In (16) and (17) the multiplication of sedenionic components is performed in accordance with (11) and Table 2. Thus the sedenionic product

$$\tilde{V}, \quad (18)$$

has the following components:

$$\theta \quad (19)$$

Note that in the sedenionic algebra the square of vector is defined as

$$\tilde{U}=\cos(\theta/2)+\boldsymbol{n}\sin(\theta/2) \quad , (20)$$

and the square of modulus of vector is

$$\tilde{U}^*. \quad (21)$$

3. SPATIAL ROTATION AND SPACE-TIME INVERSION

The rotation of sedenion $\tilde{U}^{*} = \cos(\theta/2) - \boldsymbol{n}\sin(\theta/2)$ on the angle $\tilde{U}\tilde{U}^{*} = \tilde{U}^{*}\tilde{U} = 1$ around the absolute unit vector $\tilde{V}'$ is realized by sedenion

$$\tilde{V}' = \tilde{U}^{*}\tilde{V}\tilde{U} \quad , (22)$$

and by conjugated sedenion $\tilde{V}'$:

$$\begin{aligned} \tilde{V}' &= \left[\cos(\theta/2) - \boldsymbol{n}\sin(\theta/2)\right]\left(V_0 + \vec{V}\right)\left[\cos(\theta/2) + \boldsymbol{n}\sin(\theta/2)\right] \\ &= V_0 + V\cos\theta - \boldsymbol{n}(\boldsymbol{n}\cdot V)(1-\cos\theta) - [\boldsymbol{n}\times V]\sin\theta. \end{aligned} \quad , (23)$$

with

$$\theta. \quad (24)$$

The transformed sedenion $(\hat{R}_t)$ is defined as sedenionic product

$$(\hat{R}_r), \quad (25)$$

Thus, the transformed sedenion $(\hat{R}_{tr})$ can be written as

$$\begin{aligned} \hat{R}_t\tilde{V} &= -\mathbf{e}_2\tilde{V}\mathbf{e}_2 = \overline{V}_0 - \mathbf{e}_1\overline{V}_1 + \mathbf{e}_2\overline{V}_2 - \mathbf{e}_3\overline{V}_3, \\ \hat{R}_r\tilde{V} &= -\mathbf{e}_1\tilde{V}\mathbf{e}_1 = \overline{V}_0 + \mathbf{e}_1\overline{V}_1 - \mathbf{e}_2\overline{V}_2 - \mathbf{e}_3\overline{V}_3, \\ \hat{R}_{tr}\tilde{V} &= -\mathbf{e}_3\tilde{V}\mathbf{e}_3 = \overline{V}_0 - \mathbf{e}_1\overline{V}_1 - \mathbf{e}_2\overline{V}_2 + \mathbf{e}_3\overline{V}_3. \end{aligned} \quad (26)$$

It is clearly seen that rotation does not transform the sedenion-scalar part, but the sedenionic vector $\tilde{S} = \mathbf{e}_1 ct + \mathbf{e}_2 \boldsymbol{r}$ is rotated on the angle $\tilde{L} = \operatorname{ch}\vartheta + \mathbf{e}_3\boldsymbol{m}\operatorname{sh}\vartheta$, $\tilde{S}\tilde{S} = -c^2t^2 + x^2 + y^2 + z^2$ around $\tilde{L}^{*} = \operatorname{ch}\vartheta - \mathbf{e}_3\boldsymbol{m}\operatorname{sh}\vartheta$,.

The operations of time inversion $\text{th}2\vartheta = v/c$, space inversion v and space-time inversion $\tilde{L}^{*}\tilde{L} = \tilde{L}\tilde{L}^{*} = 1$ are connected with transformations in e_1, e_2, e_3 basis and can be presented as

$$\tilde{S}' \tag{27}$$

4. SEDENIONIC LORENTZ TRANSFORMATIONS

The relativistic event four-vector can be represented in the follow sedenionic form:

$$\begin{aligned}\tilde{S}' &= \tilde{L}^{*}\,\tilde{S}\,\tilde{L} = (\text{ch}\,\vartheta - \mathbf{e}_3\text{sh}\,\vartheta\boldsymbol{m})(\mathbf{e}_1 ct + \mathbf{e}_2\boldsymbol{r})(\text{ch}\,\vartheta + \mathbf{e}_3\text{sh}\,\vartheta\boldsymbol{m}) \\ &= \mathbf{e}_1 ct\,\text{ch}2\vartheta + \mathbf{e}_1(\boldsymbol{m}\cdot\boldsymbol{r})\text{sh}2\vartheta + \mathbf{e}_2\boldsymbol{r} - \mathbf{e}_2 ct\boldsymbol{m}\,\text{sh}2\vartheta + \mathbf{e}_2(\boldsymbol{m}\cdot\boldsymbol{r})\boldsymbol{m}(1-\text{ch}2\vartheta).\end{aligned} \tag{28}$$

The square of this valueis the Lorentzinvariant

$$t' = \frac{t - xv/c^2}{\sqrt{1-v^2/c^2}},\quad x' = \frac{x - tv}{\sqrt{1-v^2/c^2}},\quad y' = y,\quad z' = z \tag{29}$$

The Lorentz transformation of event four-vector is realized by sedenions

$$\tilde{V} \tag{30}$$

where $\tilde{V}'$, $\tilde{V}' = \tilde{L}^{*}\,\tilde{V}\tilde{L}$ is velocity of motion along the absolute unit vector

$$\begin{aligned}\tilde{V}' &= (\text{ch}\,\vartheta - \mathbf{e}_{\text{tr}}\text{sh}\,\vartheta\boldsymbol{m})(\boldsymbol{V}_0 + \boldsymbol{V})(\text{ch}\,\vartheta + \mathbf{e}_{\text{tr}}\text{sh}\,\vartheta\boldsymbol{m}) \\ &= \boldsymbol{V}_0\text{ch}^2\vartheta + \mathbf{e}_{\text{tr}}\boldsymbol{V}_0\mathbf{e}_{\text{rt}}\text{sh}^2\vartheta - (\mathbf{e}_{\text{tr}}\boldsymbol{V}_0 - \boldsymbol{V}_0\mathbf{e}_{\text{tr}})\boldsymbol{m}\,\text{ch}\,\vartheta\,\text{sh}\,\vartheta + \boldsymbol{V}\text{ch}^2\vartheta \\ &\quad - \mathbf{e}_{\text{tr}}\boldsymbol{m}\boldsymbol{V}\boldsymbol{m}\mathbf{e}_{\text{tr}}\text{sh}^2\vartheta - (\mathbf{e}_{\text{tr}}\boldsymbol{m}\boldsymbol{V} - \boldsymbol{V}\boldsymbol{m}\mathbf{e}_{\text{tr}})\text{ch}\,\vartheta\,\text{sh}\,\vartheta.\end{aligned}$$

. Note that

$$\begin{aligned}\tilde{V}' &= \boldsymbol{V}_0\text{ch}^2\vartheta + \mathbf{e}_{\text{tr}}\boldsymbol{V}_0\mathbf{e}_{\text{tr}}\text{sh}^2\vartheta - (\mathbf{e}_{\text{tr}}\boldsymbol{V}_0 - \boldsymbol{V}_0\mathbf{e}_{\text{tr}})\boldsymbol{m}\,\text{ch}\,\vartheta\,\text{sh}\,\vartheta + \boldsymbol{V}\text{ch}^2\vartheta - \mathbf{e}_{\text{tr}}\boldsymbol{V}\mathbf{e}_{\text{tr}}\text{sh}^2\vartheta - 2\mathbf{e}_{\text{tr}}(\boldsymbol{m}\cdot\boldsymbol{V})\mathbf{e}_{\text{tr}}\boldsymbol{m}\text{sh}^2\vartheta \\ &\quad - (\mathbf{e}_{\text{tr}}(\boldsymbol{m}\cdot\boldsymbol{V}) - (\boldsymbol{V}\cdot\boldsymbol{m})\mathbf{e}_{\text{tr}})\text{ch}\,\vartheta\,\text{sh}\,\vartheta - (\mathbf{e}_{\text{tr}}[\boldsymbol{m}\times\boldsymbol{V}] - [\boldsymbol{V}\times\boldsymbol{m}]\mathbf{e}_{\text{tr}})\text{ch}\,\vartheta\,\text{sh}\,\vartheta.\end{aligned} \tag{31}$$

The transformed event four-vector

$$V' = V,$$
$$V'_{\mathbf{tr}} = V_{\mathbf{tr}},$$
$$V'_{\mathbf{r}} = V_{\mathbf{r}}\,\mathrm{ch}\,2\vartheta - \mathbf{e}_{\mathbf{tr}}\left(\boldsymbol{m}\cdot V_{\mathbf{t}}\right)\mathrm{sh}\,2\vartheta,$$
$$V'_{\mathbf{t}} = V_{\mathbf{t}}\,\mathrm{ch}\,2\vartheta - \mathbf{e}_{\mathbf{tr}}\left(\boldsymbol{m}\cdot V_{\mathbf{r}}\right)\mathrm{sh}\,2\vartheta,$$
$$\boldsymbol{V}' = \boldsymbol{V}\,\mathrm{ch}\,2\vartheta - \left(\boldsymbol{m}\cdot\boldsymbol{V}\right)\boldsymbol{m}\left(1-\mathrm{ch}2\vartheta\right) - \mathbf{e}_{\mathbf{tr}}\left[\boldsymbol{m}\times\boldsymbol{V}_{\mathbf{rt}}\right]\mathrm{sh}2\vartheta,$$
$$\boldsymbol{V}'_{\mathbf{tr}} = \boldsymbol{V}_{\mathbf{tr}}\,\mathrm{ch}\,2\vartheta - \left(\boldsymbol{m}\cdot\boldsymbol{V}_{\mathbf{tr}}\right)\boldsymbol{m}\left(1-\mathrm{ch}2\vartheta\right) - \mathbf{e}_{\mathbf{tr}}\left[\boldsymbol{m}\times\boldsymbol{V}\right]\mathrm{sh}2\vartheta,$$
$$\boldsymbol{V}'_{\mathbf{r}} = \boldsymbol{V}_{\mathbf{r}} + \left(\boldsymbol{m}\cdot\boldsymbol{V}_{\mathbf{r}}\right)\boldsymbol{m}\left(1-\mathrm{ch}2\vartheta\right) - \mathbf{e}_{\mathbf{tr}}V_{\mathbf{t}}\boldsymbol{m}\,\mathrm{sh}\,2\vartheta,$$
$$\boldsymbol{V}'_{\mathbf{t}} = \boldsymbol{V}_{\mathbf{t}} + \left(\boldsymbol{m}\cdot\boldsymbol{V}_{\mathbf{t}}\right)\boldsymbol{m}\left(1-\mathrm{ch}2\vartheta\right) - \mathbf{e}_{\mathbf{tr}}V_{\mathbf{r}}\boldsymbol{m}\,\mathrm{sh}\,2\vartheta.$$

is written as

$$Z_{\mathbf{t}} = z_1 + \mathbf{e}_{\mathbf{t}} z_2,$$
$$Z_{\mathbf{r}} = z_1 + \mathbf{e}_{\mathbf{r}} z_2,$$
$$Z_{\mathbf{tr}} = z_1 + \mathbf{e}_{\mathbf{tr}} z_2, \quad (32)$$

Separating the values with e_1 and e_2 we get the well known formulas for time and coordinates transformation [25] :

$$\widehat{q} = q_0\mathbf{a_0} + \mathbf{e_0}\left(q_1\mathbf{a_1} + q_2\mathbf{a_2} + q_3\mathbf{a_3}\right) \quad , (33)$$

where x is the coordinate along the $\widehat{q}_{\mathbf{t}} = q_0\mathbf{a_0} + \mathbf{e_t}\left(q_1\mathbf{a_1} + q_2\mathbf{a_2} + q_3\mathbf{a_3}\right)$ vector.

Let us also consider the Lorentz transformation of the full sedenion $\widehat{q}_{\mathbf{r}} = q_0\mathbf{a_0} + \mathbf{e_r}\left(q_1\mathbf{a_1} + q_2\mathbf{a_2} + q_3\mathbf{a_3}\right)$. The transformed sedenion $\widehat{q}_{\mathbf{tr}} = q_0\mathbf{a_0} + \mathbf{e_{tr}}\left(q_1\mathbf{a_1} + q_2\mathbf{a_2} + q_3\mathbf{a_3}\right)$ can be written as sedenionic product

$$\widehat{q}_{\mathbf{t}}. \quad (34)$$

$$\widehat{q}_{\mathbf{r}} \quad (35)$$

Rewriting the expression (35) with scalar (16) and vector (17) products we get

$$\widehat{q}_{\mathbf{tr}} \quad (36)$$

Thus, the transformed sedenion has the following components:

$$\widehat{q}_{\mathbf{t}} \qquad (37)$$

5. SUBALGEBRAS OF SPACE-TIME COMPLEX NUMBERS, QUATERNIONS AND OCTONIONS

The sedenionic basis introduced above enables constructing different types of low-dimensional hypercomplex numbers. For example, one can introduce space-time complex numbers

$$\hat{R}_{\mathbf{t}}\widehat{q}_{\mathbf{t}} = -\mathbf{e}_{\mathbf{r}}\widehat{q}_{\mathbf{t}}\mathbf{e}_{\mathbf{r}} = q_0\mathbf{a}_0 - \mathbf{e}_{\mathbf{t}}\left(q_1\mathbf{a}_1 + q_2\mathbf{a}_2 + q_3\mathbf{a}_3\right) \qquad (38)$$

where z_1 and z_2 real numbers. These values are transformed under space and time conjugation and Lorentz transformations. Moreover, we can consider the space-time quaternions, which differ in their properties with respect to the operations of the spatial and time inversion and Lorentz transformations:

$$\breve{G}_{\mathbf{t}} = G_{00} + G_{01}\mathbf{a}_1 + G_{02}\mathbf{a}_2 + G_{03}\mathbf{a}_3 + \mathbf{e}_{\mathbf{t}}G_{10} + \mathbf{e}_{\mathbf{t}}\left(G_{11}\mathbf{a}_1 + G_{12}\mathbf{a}_2 + G_{13}\mathbf{a}_3\right), \qquad (39)$$

$$\breve{G}_{\mathbf{r}} = G_{00} + G_{01}\mathbf{a}_1 + G_{02}\mathbf{a}_2 + G_{03}\mathbf{a}_3 + \mathbf{e}_{\mathbf{r}}G_{20} + \mathbf{e}_{\mathbf{r}}\left(G_{21}\mathbf{a}_1 + G_{22}\mathbf{a}_2 + G_{23}\mathbf{a}_3\right), \qquad (40)$$

$$\breve{G}_{\mathbf{tr}} = G_{00} + G_{01}\mathbf{a}_1 + G_{02}\mathbf{a}_2 + G_{03}\mathbf{a}_3 + \mathbf{e}_{\mathbf{tr}}G_{30} + \mathbf{e}_{\mathbf{tr}}\left(G_{31}\mathbf{a}_1 + G_{32}\mathbf{a}_2 + G_{33}\mathbf{a}_3\right), \qquad (41)$$

$$E^2 - c^2p^2 = m_0^2c^4 \; . \qquad (42)$$

The absolute quaternion (39) is the sum of the absolute scalar and absolute vector. It remains constant under the transformations of space and time inversion (27). Time quaternion $\hat{E} = i\hbar\frac{\partial}{\partial t}$ and $\hat{\boldsymbol{p}} = -i\hbar\nabla$, space quaternion $\hbar$ and space-time quaternion $\left(\mathbf{e}_{\mathbf{t}}E + \mathbf{e}_{\mathbf{r}}cp + i\mathbf{e}_{\mathbf{tr}}m_0c^2\right)\left(\mathbf{e}_{\mathbf{t}}E + \mathbf{e}_{\mathbf{r}}cp + i\mathbf{e}_{\mathbf{tr}}m_0c^2\right) = 0$ are transformed under inversions in accordance with the commutation rules for the basis elements e_t, e_r, e_{tr}. For example, performing the operation of time inversion (see (27)) with the quaternion $\tilde{\boldsymbol{W}}(t,\boldsymbol{r}) = \boldsymbol{W}_0(t,\boldsymbol{r}) + \boldsymbol{W}(t,\boldsymbol{r})$ we obtain the conjugated quaternion

$$\left(\mathbf{e}_t \frac{1}{c}\frac{\partial}{\partial t} - \mathbf{e}_r \nabla + \mathbf{e}_{tr} \frac{m_0 c}{\hbar}\right)\left(\mathbf{e}_t \frac{1}{c}\frac{\partial}{\partial t} - \mathbf{e}_r \nabla + \mathbf{e}_{tr} \frac{m_0 c}{\hbar}\right)\tilde{\boldsymbol{W}} = 0 \quad . \qquad (43)$$

In addition, the sedenionic basis allows one to construct various types of space-time eight-component octonions:

$$\left(\mathbf{e}_t \frac{1}{c}\frac{\partial}{\partial t} + \mathbf{e}_t \frac{ie}{\hbar c}\varphi - \mathbf{e}_r \nabla + \mathbf{e}_r \frac{ie}{\hbar c}\boldsymbol{A} + \mathbf{e}_{tr} \frac{m_0 c}{\hbar}\right)\left(\mathbf{e}_t \frac{1}{c}\frac{\partial}{\partial t} + \mathbf{e}_t \frac{ie}{\hbar c}\varphi - \mathbf{e}_r \nabla + \mathbf{e}_r \frac{ie}{\hbar c}\boldsymbol{A} + \mathbf{e}_{tr} \frac{m_0 c}{\hbar}\right)\tilde{\boldsymbol{W}} = 0 \quad , \qquad (44)$$

$$\left(\mathbf{e}_t \frac{1}{c}\frac{\partial}{\partial t} - \mathbf{e}_r \nabla + \mathbf{e}_{tr} \frac{m_0 c}{\hbar}\right)\tilde{\boldsymbol{W}} = 0 \quad , \qquad (45)$$

$$\left(\mathbf{e}_t \frac{1}{c}\frac{\partial}{\partial t} + \mathbf{e}_t \frac{ie}{\hbar c}\varphi - \mathbf{e}_r \nabla + \mathbf{e}_r \frac{ie}{\hbar c}\boldsymbol{A} + \mathbf{e}_{tr} \frac{m_0 c}{\hbar}\right)\tilde{\boldsymbol{W}} = 0 \quad . \qquad (46)$$

6. GENERALIZED SEDENIONIC EQUATIONS OF RELATIVISTIC QUANTUM MECHANICS

The wave function of free quantum particle should satisfy an equation, which is obtained from the Einstein relation for energy and momentum

$$\left(\mathbf{e}_t \frac{1}{c}\frac{\partial}{\partial t} - \mathbf{e}_r \nabla + \mathbf{e}_{tr} \frac{m_0 c}{\hbar}\right)\left(\mathbf{e}_t \frac{1}{c}\frac{\partial}{\partial t} - \mathbf{e}_r \nabla + \mathbf{e}_{tr} \frac{m_0 c}{\hbar}\right)\tilde{\boldsymbol{W}} = 0 \quad , \qquad (47)$$

by means of changing classical energy E and momentum $\tilde{\boldsymbol{W}}$ on corresponding quantum-mechanical operators:

$$\tilde{\boldsymbol{J}} \quad . \qquad (48)$$

Here c is the speed of light, m0 is the particle rest mass, $\left(\mathbf{e}_t \frac{1}{c}\frac{\partial}{\partial t} - \mathbf{e}_r \nabla + \mathbf{e}_{tr} \frac{m_0 c}{\hbar}\right)\left(\mathbf{e}_t \frac{1}{c}\frac{\partial}{\partial t} - \mathbf{e}_r \nabla + \mathbf{e}_{tr} \frac{m_0 c}{\hbar}\right)\tilde{\boldsymbol{W}} = \tilde{\boldsymbol{J}}$ is the Planck constant. In sedenion algebra the Einstein relation (47) can be written as

$$\tilde{\boldsymbol{J}} \quad . \qquad (49)$$

Let us consider the wave function in the form of space-time sedenion

$$\left(\mathbf{e}_t \frac{1}{c}\frac{\partial}{\partial t} - \mathbf{e}_r \nabla + \mathbf{e}_{tr} \frac{m_0 c}{\hbar}\right) \tilde{\boldsymbol{W}} = \tilde{\boldsymbol{I}}$$

. (50)

Then the generalized sedenionic wave equation for free particle can be written in the following symmetric form:

$$\tilde{\boldsymbol{I}}.$$

(51)

Note that for electrically charged particle in an external electromagnetic field we have the following sedenionic wave equation:

$$\tilde{\boldsymbol{W}} = \mathbf{e}_t \varphi + \mathbf{e}_r \boldsymbol{A}.$$

(52)

This equation describes the particles with spin 1/2 in an external electromagnetic field [18] [21] .

There is a special class of particles described by the first-order wave equation [26] . For these particles the sedenionic Dirac-like wave equation has the following form:

$$\tilde{\boldsymbol{J}} = -\mathbf{e}_t 4\pi\rho - \mathbf{e}_r \frac{4\pi}{c} \boldsymbol{j}.$$

(53)

In fact, this equation describes the special quantum field with zero field strengths [19] . Analogously the electrically charged particle interacting with external electromagnetic field is described by the following sedenionic first-order wave equation:

$$\left(\mathbf{e}_t \frac{1}{c}\frac{\partial}{\partial t} - \mathbf{e}_r \nabla\right)\left(\mathbf{e}_t \frac{1}{c}\frac{\partial}{\partial t} - \mathbf{e}_r \nabla\right)\left(\mathbf{e}_t \varphi + \mathbf{e}_r \boldsymbol{A}\right) = -\mathbf{e}_t 4\pi\rho - \mathbf{e}_r \frac{4\pi}{c} \boldsymbol{j}.$$

(54)

This equation also describes particles with spin 1/2 in an external electromagnetic field [19] .

7. GENERALIZED SEDENIONIC EQUATIONS FOR MASSIVE FIELD

The generalized sedenionic wave equation

$$\left(\mathbf{e}_t \frac{1}{c}\frac{\partial}{\partial t} - \mathbf{e}_r \nabla\right)\left(\mathbf{e}_t \varphi + \mathbf{e}_r \boldsymbol{A}\right) = -\frac{1}{c}\frac{\partial \varphi}{\partial t} + \mathbf{e}_{tr} \frac{1}{c}\frac{\partial \boldsymbol{A}}{\partial t} + \mathbf{e}_{tr} \nabla \varphi + (\nabla \cdot \boldsymbol{A}) + [\nabla \times \boldsymbol{A}],$$

(55)

enables another interpretation. It can be considered as the equation for the force massive field [27] . In this case the parameter m_0 is the mass of quantum of field and $E = -\frac{1}{c}\frac{\partial A}{\partial t} - \nabla\varphi,$ $H = [\nabla \times A].$ is field potential. Considering the phenomenological source of field $f = -\frac{1}{c}\frac{\partial \varphi}{\partial t} + (\nabla \cdot A)$ we can propose the following nonhomogeneous wave equation for the field potential:

$$\left(\mathbf{e}_t \frac{1}{c}\frac{\partial}{\partial t} - \mathbf{e}_r \nabla\right)\left(\mathbf{e}_t \varphi + \mathbf{e}_r A\right) = -\mathbf{e}_{tr} E + H \quad . (56)$$

Seemingly this equation describes the baryon (strong) field [23] [24] and $\left(\mathbf{e}_t \frac{1}{c}\frac{\partial}{\partial t} - \mathbf{e}_r \nabla\right)\left(-\mathbf{e}_{tr} E + H\right) = -\mathbf{e}_t 4\pi\rho - \mathbf{e}_r \frac{4\pi}{c} j$ is baryon current. On the other hand, corresponding nonhomogeneous first-order equation

$$\mathbf{e}_r \frac{1}{c}\frac{\partial E}{\partial t} + \mathbf{e}_t (\nabla \cdot E) + \mathbf{e}_t [\nabla \times E] + \mathbf{e}_t \frac{1}{c}\frac{\partial H}{\partial t} - \mathbf{e}_r (\nabla \cdot H) - \mathbf{e}_r [\nabla \times H] = -\mathbf{e}_t 4\pi\rho - \mathbf{e}_r \frac{4\pi}{c} j \quad , (57)$$

describes the lepton (weak) field, where

$$\begin{aligned}
&\mathbf{e}_t (\nabla \cdot E) = -\mathbf{e}_t 4\pi\rho, && (\text{time scalar part}) \\
&\mathbf{e}_r [\nabla \times H] = \mathbf{e}_r \frac{1}{c}\frac{\partial E}{\partial t} + \mathbf{e}_r \frac{4\pi}{c} j, && (\text{space vector part}) \\
&\mathbf{e}_t [\nabla \times E] = -\mathbf{e}_t \frac{1}{c}\frac{\partial H}{\partial t}, && (\text{time vector part}) \\
&\mathbf{e}_r (\nabla \cdot H) = 0. && (\text{space scalar part})
\end{aligned}$$

is a lepton current [23] [24] .

8. GENERALIZED SEDENIONIC EQUATIONS FOR MASSLESS FIELD

In the special case, when the mass of quantum m_0 is equal to zero, the Equation (56) coincides with the equation for electromagnetic field. Indeed, choosing the potential as

$$\left(\mathbf{e}_t \frac{1}{c}\frac{\partial}{\partial t} - \mathbf{e}_r \nabla\right)\tilde{\boldsymbol{W}}_\nu = 0$$
, (58)

and the source of field as

$$\left(\mathbf{e}_t \frac{1}{c}\frac{\partial}{\partial t} - \mathbf{e}_r \nabla\right)\tilde{\boldsymbol{W}}_\nu = \tilde{\boldsymbol{I}}_\nu$$
, (59)

we obtain the following wave equation:

$$\tilde{\boldsymbol{I}}_\nu \ . \qquad (60)$$

After the action of the first operator in the left-hand side of Equation (60) we obtain

$$\tilde{\boldsymbol{I}}_\nu = 4\pi\sigma_\nu \ . \qquad (61)$$

In sedenionic algebra the electric and magnetic fields are defined as

$$\sigma_\nu \qquad (62)$$

Besides we can define the scalar field

$$\tilde{\boldsymbol{W}}_\nu \ . \qquad (63)$$

Assuming electric charge conservation the scalar field f can be chosen equal to zero, that coincides with Lorentz gauge condition [22] . In Lorentz gauge we can rewrite the expression (61) as

$$\tilde{\boldsymbol{W}}_\nu = i\mathbf{e}_t\varphi_\nu + \mathbf{e}_r\boldsymbol{A}_\nu \ . \qquad (64)$$

Then the wave Equation (60) can be represented in the following form:

$$\left(\mathbf{e}_t \frac{1}{c}\frac{\partial}{\partial t} - \mathbf{e}_r \nabla\right)\left(\mathbf{e}_t\varphi_\nu + \mathbf{e}_r\boldsymbol{A}_\nu\right) = 4\pi\sigma_\nu \ . \qquad (65)$$

Performing sedenionic multiplication in the left-hand side of Equation (65) we get

$$f_\nu \ . \qquad (66)$$

Separating space-time values we obtain the system of equations in the following form:

$$f_v = 4\pi\sigma_v \tag{67}$$

The system (67) coincides with the Maxwell equations.

Among the solutions of the homogeneous sedeonic wave equation of electromagnetic field (60) there is a special class that satisfies the sedeonic first-order equation of the following form [22] :

$$\sigma_v = q_v \delta(\boldsymbol{r}) \quad . \tag{68}$$

This equation describes the free neutrino field. On the other hand, let us consider the nonhomogeneous equation of neutrino field

$$W_{v1v2} = \frac{1}{8\pi}\int f_{v1} f_{v2} \mathrm{d}V \quad , \tag{69}$$

where $W_{v1v2} = 2\pi q_{ve1} q_{ve2} \delta(\boldsymbol{R})$ is phenomenological source. We choose the scalar source in the form

$$\mathbf{a}'_{\mathbf{n}} \; , \tag{70}$$

where $\mathbf{e}'_{\mathbf{n}}$ is the density of neutrino charge. Choosing the potential $\mathbf{a}'_{\mathbf{n}} = i\mathbf{a}_{\mathbf{n}}$ in the form (58):

$$\mathbf{e}'_{\mathbf{n}} = i\mathbf{e}_{\mathbf{n}} \; , \tag{71}$$

we obtain following nonhomogeneous equation for the neutrino field:

$$\left(\mathbf{e}_{\mathrm{t}} \frac{1}{c}\frac{\partial}{\partial t} - \mathbf{e}_{\mathrm{r}}\vec{\nabla}\right)\left(\mathbf{e}_{\mathrm{t}}\varphi_v + \mathbf{e}_{\mathrm{r}}\vec{A}_v\right) = 4\pi\sigma_v \quad . \tag{72}$$

It follows that in this case only scalar field strength f_v (see (63)) is nonzero:

$$f_v = 4\pi\sigma_v \; . \tag{73}$$

The density of neutrino charge for point source is equal

$$\sigma_v = q_v \delta(\vec{r}) \quad (74)$$

where q_v is point neutrino charge. Then the interaction energy of two point neutrino charges can be represented as follows:

$$W_{v1v2} = \frac{1}{8\pi}\int f_{v1} f_{v2} \mathrm{d}V \quad (75)$$

Substituting (73) and (74), we obtain

$$W_{v1v2} = 2\pi q_{v1} q_{v2} \delta(\vec{R}) \quad (76)$$

where $\vec{R}$ is the vector of distance between first and second charges.

9. DISCUSSION

The algebra of sedenions proposed in this article is the anticommutative associative space-time Clifford algebra. The sedenionic basis elements a_n are responsible for the spatial rotation, while the elements e_n are responsible for the space-time inversions. Mathematically, these two bases are equivalent, and the different physical properties attributed to them are an important physical essence of our sedenionic hypothesis.

In contrast to the previously discussed sedeonic algebra [20] -[23] , which uses the multiplication rules of basic elements $\mathbf{a}'_{\mathbf{n}}$ and $\mathbf{e}'_{\mathbf{n}}$ proposed by A. Macfarlane [28] , the multiplication rules for sedenionic basis elements a_n and e_n coincide with the rules for quaternion units introduced by W. R. Hamilton [29] . There is a close connection between these two basses. The transition from the sedeonic basis to sedenionic basis is performed by following replacement:

$$\mathbf{a}'_{\mathbf{n}} = i\mathbf{a}_{\mathbf{n}},$$

$$\mathbf{e}'_{\mathbf{n}} = i\mathbf{e}_{\mathbf{n}}.$$

There is one disadvantage of sedenions connected with the fact that the square of the vector is a negative value. However, on the other side the sedenionic rules of cross-multiplying do not contain the imaginary unit and this leads to the some simplifications in the calculations. But of course, the physical results do not depend on the choice of algebra, so these two algebras are equivalent.

10. CONCLUSION

Thus, in this paper we presented the sixteen-component hypercomplex values sedenions, generating associative noncommutative space-time algebra. We considered the generalization of the relativistic quantum mechanics and theory of massive and massless fields based on hypercomplex scalar-vector wave functions and sedenionic space-time operators.

ACKNOWLEDGEMENTS

The authors are very thankful to G. V. Mironova for kind assistance and moral support.

REFERENCES

1. Adler, S.L. (1995) Quaternionic Quantum Mechanics and Quantum Fields. Oxford University Press, New York.
2. Imaeda, K. (1976) A New Formulation of Classical Electrodynamics. Nuovocimento, 32, 138-162.
3. Majernik, V. (1999) Quaternionic Formulation of the Classical Fields. Advances in Applied Clifford Algebras, 9, 119- 130.
4. Davies, A.J. (1990) Quaternionic Dirac Equation. Physical Review D, 41, 2628-2630.
5. Schwartz, C. (2006) Relativistic Quaternionic Wave Equation. Journal of Mathematical Physics, 47, Article ID: 122301.
6. Liu, Y.-F. (2002) Triality, Biquaternion and Vector Representation of the Dirac Equation. Advances in Applied Clifford Algebras, 12, 109-124.
7. Gogberashvili, M. (2006) Octonionic Electrodynamics. Journal of Physics A: Mathematics in General, 39, 7099-7104.

8. Gamba, A. (1998) Maxwell's Equations in Octonion Form. NuovoCimento A, 111, 293-302.

9. Tolan, T., Özdas, K. and Tanisli, M. (2006) Reformulation of Electromagnetism with Octonions. Nuovo Cimento B, 121, 43-55.

10. Gogberashvili, M. (2006) Octonionic Version of Dirac Equations. International Journal of Modern Physics A, 21, 3513-3523.

11. De Leo, S. and Abdel-Khalek, K. (1996) Octonionic Dirac Equation. Progress in Theoretical Physics, 96, 833-846.

12. Dickson, L.E. (1919) On Quaternions and Their Generalization and the History of the Eight Square Theorem. Annals of Mathematics, 20, 155-171.

13. Imaeda, K. and Imaeda, M. (2000) Sedenions: Algebra and Analysis. Applied Mathematics and Computations, 115, 77- 88.

14. Carmody, K. (1988) Circular and Hyperbolic Quaternions, Octonions, and Sedenions. Applied Mathematics and Computations, 28, 47-72.

15. Carmody, K. (1997) Circular and Hyperbolic Quaternions, Octonions, and Sedenions—Further Results. Applied Mathematics and Computations, 84, 27-47.

16. Köplinger, J. (2006) Dirac Equation on Hyperbolic Octonions. Applied Mathematics and Computations, 182, 443-446.

17. Mironov, V.L. and Mironov, S.V. (2009) Octonic Representation of Electromagnetic Field Equations. Journal of Mathematical Physics, 50, Article ID: 012901.

18. Mironov, V.L. and Mironov, S.V. (2009) Octonic Second-Order Equations of Relativistic Quantum Mechanics. Journal of Mathematical Physics, 50, Article ID: 012302.

19. Mironov, V.L. and Mironov, S.V. (2009) Octonic First-Order Equations of Relativistic Quantum Mechanics. International Journal of Modern Physics A, 24, 4157-4167.

20. Mironov, V.L. and Mironov, S.V. (2011) Noncommutative Sedeons and Their Application in Field Theory. e-Print.

21. Mironov, V.L. and Mironov, S.V. (2013) Reformulation of Relativistic Quantum Mechanics Equations with Non- Commutative Sedeons. Applied Mathematics, 4, 53-60.

22. Mironov, V.L. and Mironov, S.V. (2014) Sedeonic Equations of Gravitoelectromagnetism. Journal of Modern Physics, 5, 917-927.

23. Mironov, S.V. and Mironov, V.L. (2014) Sedeonic Equations of Massive Fields. International Journal of Theoretical Physics, e-Print.

24. Mironov, S.V. and Mironov, V.L. (2014) Space-Time Sedeons and Their Application in Relativistic Quantum Mechanics and Field Theory. Institute for Physics of Microstructures RAS, Nizhny Novgorod.

25. Landau, L.D. and Lifshits, E.M. (1975) Classical Theory of Fields. 4th Edition, Pergamon Press, New York.

26. Dirac, P.A.M. (1958) The Principles of Quantum Mechanics. Clarendon Press, Oxford.

27. Cafaro, C. and Ali, S.A. (2006) The Spacetime Algebra Approach to Massive Classical Electrodynamics with Magnetic Monopoles. Advances in Applied Clifford Algebras, 17, 23-36.

28. Macfarlane, A. (1900) Hyperbolic Quaternions. Proceedings of the Royal Society at Edinburgh, 1899-1900 Session, 169-181.

29. Hamilton, W.R. (1853) Lectures on Quaternions. Royal Irish Academy, Dublin.

12 CHAPTER

SPATIAL MEANING OF QUANTUM MECHANICS

Yougang Feng

College of Science, Guizhou University, Guiyang, China

ABSTRACT

We provide theoretical evidence for that remains far from clear in Copenhagen interpretation, and then try to make it further complete. Uncertainty relations are proved to be the intrinsic attributes of the position-momentum space and the time-energy space. A theoretical evidence for the probabilistic interpretation is given. Different meanings of the wave-particle duality for the photons and for the electron are discussed.

Keywords: Nonlocality, Uncertainty, Quantization, Probabilistic Interpretation

1. INTRODUCTION

Quantum mechanics arises from the long and successful efforts of scientists to explain experimental phenomena such as the radiation law and the photoelectric effect, for which cannot account classical mechanics. Classical physics always represents our world in the relatively graphical and intuitive terms, greatly different from quantum physics, which is strong on mathematical rigor but light on physical intuition. The quantum properties of microscopic matter imply a renunciation of the possibility of their being understood in the customary imaginative sense, which is difficult to be admitted or embarrassed to be accepted for some classical physicists. The discussions about the principle of nonlocality of the material particles have never been stopped since quantum theory was developed in the 1920s [1] [2] . The argument relates mainly to two queries: uncertainty principle and probabilistic interpretation of wave functions. In the following we will review both issues in turn, and the first is uncertainty principle. All we should do is to provide added evidence and derivation for those remain far from clear in Copenhagen interpretation, and then make it further complete.

In general, a prevailing perception from our daily experience is that the laws of physics should be completely objective, and they should not depend in any way on what dimensions and mass of a moving object has. Namely, there should be uniform laws of physics to describe moving objects within a common frame of reference. Einstein suggested that the particle's motion should be deterministic and realistic like classical objects rather than probabilistic. Copenhagen interpretation is regarded as anti-realistic because Einstein's constant researching and sever criticism. Here we first think of an example of the interrelation between Newton's mechanism and Einstein's special theory of relativity. The physics of Newton provides an ultimate inertial frame of reference with both absolute space and time, in which all motion could be in principle measured accurately. Einstein's special theory of relativity revised Newton's point of view and gave us new concepts—there are only relativistic space and time but absolute ones. Lorentz transformation can be simplified into Galiliean transformation only if an object's velocity is far smaller than light speed, which leads to the fact that an absolute space and an absolute time come true. Some authors mentioned without rigorous proof that the dimensions and mass of a macroscopic object are too great to consider uncertainty principle, that is, Plank's constant is assigned as zero, and then the object's motions are to be thought of as deterministic [3] . An essential problem is whether or not the uncertainty principle is intrinsically applicable to classical physics, whatever the dimensions and masses. If the principle satisfies classical physics, up to now, why cannot we get a derivation of the principle starting from the laws of classical physics? How should we derive the principle without the means of quantum mechanics if the principle is an objectively general law? So far, we have known that the noncommuting properties of the operators of observables result in uncertainty relations, and these operators only exist in a complex space where the wave equation of quantum mechanics works. In contrast, the object's motion only undergoes in a real space where classical physics works. For this reason, the uncertainty principle is usually viewed as a crucial base for quantum mechanics rather than classical physics. In a practical sense, we can say there is no uncertainty principle and there is no probabilistic interpretation. In Einstein's opinion it seems that a moving object in a real space does not obey the principle unless the principle is an attribute of the real space rather than a purely algebraic description of nature [4] . We think that there are no space and time, there are no physical laws. We should not ignore the properties of the space and time, which directly affect our observation and knowledge on nature. In the second section of this paper we will prove that uncertainty relation is an intrinsic attribute of a real space independent of any moving object such that the uncertainty principle is available to all objects, no matter what dimensions and masses they will have. Classical physics is only a peculiar consequence of the elimination of these attributes, like Newton's physics is a result of neglecting the relativistic effect. As a result, quantum formalism is only taken as a good approximate description of microscopic particles rather than a real motion law. For example, the arrival of an electron at a detector in double-slit interference experiment can be rationally considered as one of our expecting events, just like there appeared to be a good weather sometimes on our prediction. Our derivation will also confirm a fact that there is not any evidence so far showing that the collapse of the wave functions is a real physical phenomenon.

We review the probabilistic interpretation in the following. Its kernel is the principle of nonlocality for the matter particles, which has been tested and verified by experiments supported by modern technique and devices. It is said that Bell's inequality is a way to distinguish between quantum mechanics and classical mechanics for the behavior of particles such as photons. The results of the experiments using Bell's inequality have now come down in favor of quantum mechanics [5] -[7] . In fact, it is impossible to show classical physics phenomenon for a quantum system, no matter what a kind of experiment is taken. Because when we make an experiment not only the quantum system but also the measuring apparatus are placed in a position-momentum space and a timeenergy space, both system and the apparatus are absolutely affected by the uncertainty principle belonging to the physical nature of these spaces. For this reason, Bell's inequality is only an experiment showing once more the fundamental phenomena for quantum entities. Multiphoton experiments tests of quantum nonlocality show that Bell's theorem is not sufficiently theoretical argument [8] [9] . Born's probabilistic interpretation itself is pure guesswork but a result of theoretical evolution [10] . It is the guesswork that leads to occurrence of some alternatives to Copenhagen interpretation such as the ever popular "Many-world" interpretation of quantum mechanics, "Hidden-variable" theory, "Consistent-histories" interpretation, and "Decoherenc" mechanism [11] -[14] . Furthermore, there are a series of oppositional voice including the Einstein's EPR paradox and the Schrödinger's cat paradox, which directly blame Born's interpretation [15][16] . We think that there should be a theoretical proof for the nonlocality principle so as to decrease the disputation in the future. Born's guesswork based on two facts: The first is Heisenberg uncertainty principle; the second is an idea of Einstein's interpreting the square of the optical amplitudes as probability density for the occurrence of photons. In our opinion, the wave-particle duality of light is most helpful in visualizing the nature of the duality, which directly leads to their probabilistic distribution. It is clear that there is no photon concept prior to Plank's quantum theory and Einstein's photoelectric effect approach. In a homogeneous medium a free electromagnetic field also is called radiation field due to free of charges and currents, and shows a wave character that is called light wave, and the wave functions are just the field's vector potential. The field energies are discretized by Plank's theory, corresponding to quanta called photons by Einstein. Dirac first recognized that the dynamical properties of light wave functions are equivalent to the dynamical ones of a set of oscillators; they are just the same one looked at from two different points of view. One oscillator associated with each independent wave function state, which enables a function of the wave and corpuscular theories of light to be effected [17] . Therefore, we can say the radiation field is characterized by the simply harmonic wave, and the wave quantization results in photons. Two types of particles similar to photons from the quantization are phonons and magnons in solid state physics; the former results from the lattice wave quantization, and the latter are the products of the spin wave quantization [18] . It should be emphasized that there is no wave and there is no the relevant quantization. Alternatively, the quantization of a temperature field is not physically accepted. Hence, the terms of the wave quantization have more definite meaning than the terms of the field quantization. In contrast with the fact that Maxwell's field equation is prior to the photon concept, the electron concept is prior to Schrödinger wave equation. Schrödinger imagined an electron as a wave with zero wavelength, obeying Hamilton's principle, and the wave was directly influenced only by its immediate surroundings [19] . In order to make the energies discrete, he treated the Hamiltonian as an operator, and the relevant energies are amenable to Plank theory. Meanwhile, the wave functions determined by his equation turn into the eigenvector in complex space. Before further discussion about the wave-particle duality it is of enlightening meaning to reassert the concept of electric energy. Electric potential energy accounts for work on an electron done by electric force and gives added insight regarding energy and energy transformation without the necessity of dealing with the force directly. The energy does not belong to the electron alone, but both the electron and the electric field. For

Schrödinger equation, it is an inappropriate perspective to separate the electron from the Hamiltonian and assign the energies only to the electron, regardless of the electric field. A space consisting of electron and electric field is different from vacuum, and such a space may be called a field. As such, it is the space (or the field) rather than the electron exhibits a wave phenomenon. The wave like light wave can be made quantization to give particles, none of which is the electron that is a material particle. It can be understood since de Broglie's work antedated Schrödinger equation and Dirac radiation field theory, and his postulate was not able to ascribe the wave properties to the space containing electrons. It is the continuous motion of the electron that causes the wave. It seems that Schrödinger did not perceive that the wave originates in the electron's motion, and the wave functions can change with time even if for a steady system. The wave motion requires each wave function to vibrate at its own site. The wave hence is a disturbance of the wave functions in the space. A wave function at a position transfers its physical state in this way to the wave function at the next position. The wave as a whole is to be made quantization to produce particles to be like photons. Even though Schrödinger equation for single electron gives perfectly accurate solution, there is still more or less deficiency of dynamical characteristics for the wave functions. His time-dependent wave equation is a second-order differential in the three spatial coordinates but only a first-order differential in time. Actually, the wave motion is a classical phenomenon satisfying a classical dynamical equation with a second-order differential in time. In addition, the wave function of a system of N electrons is a function in a 3N-dimensional configuration space, in which the Born's probabilistic interpretation cannot be used. It is difficult to imagine that a physical reality exists in such an abstract multidimensional configuration space. The double-slit experiment is a successful example of the probabilistic interpretation [14] [20] . The behavior of an electron in the space between the slit and the detector shows the characteristics of a free electron. For these reasons, in the third section of this paper, the wave functions for the space involving a free electron are obtained by means of a dynamical equation. Then, the wave quantization for such space will be made, which will give a theoretical evidence of the probabilistic interpretation. In the fourth section we discuss different meanings of wave-particles for photons and for electron. The relativistic effect will not be discussed in this paper.

2. UNCERTAINTY RELATIONS AS ATTRIBUTES OF SPACES

In a 3-dimensional flat real space of coordinates x, y, and z there is a cotangent space with its natural bases, $\mathrm{d}x$, $\mathrm{d}y$, and $\mathrm{d}z$, together with which there is a tangent space with natural bases $\partial/\partial x$, $\partial/\partial y$, and $\partial/\partial z$. The two vector spaces are dual to each other in terms of mathematics, and their dual bases obey the following formula [21] :

$$\frac{\partial}{\partial x}\mathrm{d}x = 1, \quad \frac{\partial}{\partial y}\mathrm{d}y = 1, \quad \frac{\partial}{\partial z}\mathrm{d}z = 1 \tag{1}$$

where $\frac{\partial}{\partial x}$, $\frac{\partial}{\partial y}$, and $\frac{\partial}{\partial z}$ are also known as partial differential operators. Equation (1) is a restriction relation between dual bases. Suppose that there is a linear momentum space with its bases $C\frac{\partial}{\partial x}$, $C\frac{\partial}{\partial y}$, and $C\frac{\partial}{\partial z}$, where C is a proportional constant and makes these bases have a momentum dimension. In physical sense, $C\frac{\partial}{\partial x}$ is parallel to $\frac{\partial}{\partial x}$ because a velocity associated with $C\frac{\partial}{\partial x}$ is along the tangent direction. At the same time, $C\frac{\partial}{\partial y}$ and $C\frac{\partial}{\partial z}$ are parallel to $\frac{\partial}{\partial y}$ and $\frac{\partial}{\partial z}$, respectively. From Equation (1), we get further

$$C\frac{\partial}{\partial x}\mathrm{d}x = C, \quad C\frac{\partial}{\partial y}\mathrm{d}y = C, \quad C\frac{\partial}{\partial z}\mathrm{d}z = C \tag{2}$$

According to differential geometry [21] , a complex structure on a vector space tantamount to multiplying the basis of the vector space by an imaginary number i, which represents the square root of -1. Such procedure is called map in mathematics sense. Thus, a complex basis corresponding to $C\frac{\partial}{\partial x}$ is $iC\frac{\partial}{\partial x}$, which complex conjugational basis is $-iC\frac{\partial}{\partial x}$.

Compared with the momentum operator $-i\hbar\frac{\partial}{\partial x}$ of quantum mechanics, we right now understand the constant C does be Plank's constant due to the map is one-to-one. Equation (2) then becomes

$$\hbar\frac{\partial}{\partial x}dx = \hbar, \quad \hbar\frac{\partial}{\partial y}dy = \hbar, \quad \hbar\frac{\partial}{\partial z}dz = \hbar \tag{3}$$

A natural basis of a vector space is different from a unit vector. There is no a specific length definition for a natural basis in mathematics, its length is uncertain. In the special theory of relativity the length of a basis $\mathrm{d}x$ is considered as a distance or an interval, represented as $\mathrm{d}X$, and $\mathrm{d}X^2 = g\mathrm{d}x^2$, where g is Riemannian metric [22] . In a region where Galilea transformation is valid the Riemannian metric equals to a unit of length. In terms of mathematics, the length of $\mathrm{d}x$ can be written as $|\mathrm{d}x|$, which is its module, so $\mathrm{d}X^2 = |\mathrm{d}x|^2$. The $\mathrm{d}X$

is a measurable quantity equivalent to Δx of quantum mechanics. We then have

$$\Delta x^2 = (\Delta x)^2 = |\mathrm{d}x|^2 \tag{4}$$

Likewise,

$$\Delta p_x^2 = (\Delta p_x)^2 = \left|\hbar\frac{\partial}{\partial x}\right|^2 \tag{5}$$

Using Equations (3)-(5), we get

$$\Delta x^2 \cdot \Delta p_x^2 = \hbar^2 \tag{6}$$

Clearly, we do not know the individual values of Δx^2 and Δp_x^2, except their mutual restricting relation Equation (6). We can consider the lengths of the bases at a particular point $(\langle x\rangle,\langle p_x\rangle)$ in a 2-dimensional position-momentum space (called a phase space in statistical mechanics), where $\langle x\rangle$ and $\langle p_x\rangle$ are statistical average values of x and p_x, respectively. The starting point of the basis $\mathrm{d}x$ positions at the $\langle x\rangle$, and its end point arrives at the x, and the length is Δx. The basis $\hbar\frac{\partial}{\partial x}$ starts from the $\langle p_x\rangle$, and ends at the p_x, its length is Δp_x. In a measurable sense, Δx^2 and Δp_x^2 can be expressed by

$$\Delta x^2=\left\langle\left(x-\langle x\rangle\right)^2\right\rangle \tag{7}$$

$$\Delta p_x^2=\left\langle\left(p_x-\langle p_x\rangle\right)^2\right\rangle \tag{8}$$

where the symbol $\langle\cdots\rangle$ represents making average calculation. Equations (6)-(8) manifest uncertainty relation about position and momentum. With the same reason, we can get other uncertainty relations about Δy^2, Δp_y^2, Δz^2, and Δp_z^2. The above derivation is as simple as the Heisenberg's derivation in quantum mechanics.

Although the time-energy uncertainty relation is consistent with experiment, it is not derived from the noncommuting property of relative operators since there is no time operator in quantum theory. We think that this relation should be an intrinsic attribute for the time-energy space. When mentioned the space-time Feynman pointed out [23] : "We have discovered four quantities which transform like x, y, z, and t, and which we call the four-vector-momentum, ..., it can be represented on a space-time diagram of a moving particle as an "arrow" tangent to the path. This is an arrow has a time component equal to the energy". According Feynman, energy and time are dual to each other. At first, we get a dual relation for a cotangent basis $\mathrm{d}t$ and tangent basis $\frac{\partial}{\partial t}$, and $\frac{\partial}{\partial t}\mathrm{d}t = 1$. An energy basis can be represented by $\hbar\frac{\partial}{\partial t}$, which correspondingly mapping complex basis is $i\hbar\frac{\partial}{\partial t}$, called energy operator in terms of quantum mechanics. We then get $\hbar\frac{\partial}{\partial t}\mathrm{d}t = \hbar$. At a point $(\langle E\rangle,\langle t\rangle)$ in a 2-dimensional time-energy space, wc obtain the uncertainty relation: $\Delta E^2 \cdot \Delta t^2 = \hbar^2$, where

$$\Delta E^2 = \left|\hbar\frac{\partial}{\partial t}\right|^2 = \left\langle\left(E-\langle E\rangle\right)^2\right\rangle \tag{9}$$

$$\Delta t^2 = |\mathrm{d}t|^2 = \left\langle\left(t-\langle t\rangle\right)^2\right\rangle \tag{10}$$

The relative Riemannian metric equals one unit length due to the flat space.

The above derivation indicates that uncertainty relations are really the attributes of the real space, but belongs to microscopic particles. When we want to describe an object's motion, no matter whether the object is macroscopic or microscopic, we need to set up the space-time coordinate system in a reference frame, which means that the uncertainty relations are available for both classical physics and quantum physics, and affect a whole process of the motion. It is the intrinsic attributes that limit our ability to reach an underlying physical reality of the particles. Quantum mechanics is only an artificial description resulting from our perception of microscopic particles behavior. Just as Aage Petersen, Bohr's assistant pointed out [3] : "There is no quantum world. There is an abstract quantum physical description. It is wrong to think that the test of physics is to find out how nature is. Physics concerns what we can say about nature."

3. A Theoretical Evidence of Born's Guesswork

We try to discuss a wave that propagates freely. Our demonstration starts in classical fashion of the wave available to Newton's law. For brevity, we consider a case of one dimensional configuration space. The points on x-axis have the same nearest neighbor space denoted by a number ε, which is sufficiently small but zero. Such division resembles the approach used in the path integral [24] , which mathematical significance is that the space is a metric space, in which two distinct points can always be surrounded by disjoint domains [25] . The identical interval is due to the space is homogeneous. A space contains only single free electron, and there is not any external force. As Kubo's theorem states that the fluctuation in the space will decay [26] , but vanish due to uncertainty principle. Thus, the fluctuation force is weak. Let $\psi(x,t)$ denote a wave function at position x at time t, the domain of x be $-\infty < x < +\infty$. $\psi(x-\varepsilon,t)$ and $\psi(x+\varepsilon,t)$ are the nearest neighbors of $\psi(x,t)$, respectively. We get

$$\begin{aligned}\mu\frac{\mathrm{d}^2\psi(x,t)}{\mathrm{d}^2 t} &= C\cdot\left[\psi(x+\varepsilon,t)-\psi(x,t)\right]+C\cdot\left[\psi(x-\varepsilon,t)-\psi(x,t)\right]\\ &= C\cdot\left[\psi(x+\varepsilon,t)+\psi(x-\varepsilon,t)-2\psi(x,t)\right].\end{aligned} \tag{11}$$

where μ is equivalent mass and C is a force constant. This is a difference equation in the deviation of wavefunctions, and the fluctuation force is linear in the deviation of the function values. The equation of motion is

$$\frac{\mathrm{d}^2\psi(x,t)}{\mathrm{d}t^2} = -\omega^2\psi(x,t) \tag{12}$$

where angular frequency is ω. By Equations (11) and (12), we have

$$-\mu\omega^2\psi(x,t) = C\cdot\left[\psi(x+\varepsilon,t)+\psi(x-\varepsilon,t)-2\psi(x,t)\right] \tag{13}$$

Equation (13) has a traveling-wave solution in the form

$$\psi(x \pm \varepsilon, t) = u\exp\left[i(x \pm \varepsilon)k\right]\exp(-iwt) \tag{14}$$

where i is imaginary, the k is wave vector, u is a real constant. Inserting Equation (14) in Equation (13) yields

$$\omega^2 \mu = -C \cdot \left[\exp(ik\varepsilon) + \exp(-ik\varepsilon) - 2\right] \tag{15}$$

Using identity $2\cos k\varepsilon = \exp(ik\varepsilon) + \exp(-ik\varepsilon)$, we obtain a dispersion relation between ω and k

$$\omega^2 = (2C/\mu)(1 - \cos k\varepsilon) \tag{16}$$

Since the frequency ω depends on the magnitude of the wave vector k, it is rewritten as ω_k, and $\omega_k = \omega_{-k}$. The wave function then becomes $\psi_k(x,t)$ for a special mode k. The period of the function $\cos k\varepsilon$ is determined by the total number N of possible vibration modes, and the range of underlying values of k is specified by

$$-\pi < k\varepsilon \le \pi \tag{17}$$

We put $L = (N-1)\varepsilon$, which means that an identical mode repeatedly appears at the same time every propagating distance L relative to a periodic boundary condition. The following discussion is limited to the region L because of the periodic space. Therefore, the solution of Equation (12) is

$$\psi_k(x,t) = u\exp\left[i\left(kx - \omega_k t\right)\right] \tag{18}$$

The de Broglie relation gives the momentum $p = k\hbar$, inserting which in Equation (18), we get

$$\psi_k(x,t) = u\exp\left[i\left(\frac{px}{\hbar} - \omega_k t\right)\right] \tag{19}$$

It is well known the wave function of an eigenvalue equation in the momentum representation for free particle is

$$\psi(x) = \exp\left(\frac{ipx}{\hbar}\right) \tag{20}$$

Clearly, Equation (20) is just a special form of $\psi_k(x,t)$ at $t = 0$ in Equation (19). Note that there is only an energy which is just the electron's one, and the position x is the where the electron possibly appears. If we substitute the electron kinetic energy of $p^2/(2m_e) = \hbar\omega_k$ in Equation (19), we obtain

$$\psi_k(x,t) = u\exp\left(\frac{i}{\hbar}px - \frac{ip^2}{\hbar 2m_e}t\right) \tag{21}$$

where m_e is the electron mass. This is consistent with the function given by Feynman in the path integral [24] , which is explained in terms of the probability amplitude for free electron to arrive at the point x at the time t. It is certain that the wave functions of the free electron in quantum mechanics and path integral satisfy the equation of motion, which justifies that the space involving a free electron shows wave property. Since the wave motion is analogous to the simply harmonic oscillation, the subsequent wave quantization is expected. We make Fourier transformation for $\psi(x,t)$

$$\psi(x,t)=\sum_k \psi_k(x,t)=\frac{1}{\sqrt{N\mu}}\sum_k Q_k(t)\exp(ikx) \tag{22}$$

Let $P_k = \mathrm{d}Q_k/\mathrm{d}t$. The following treatment is familiar in quantum mechanics [27] . We introduce operators

$$\hat{\alpha}_k=\left(\frac{\mu\omega_k}{2\hbar}\right)^{1/2}\left(\hat{Q}_k+\frac{i}{\mu\omega_k}\hat{P}_{-k}\right),\quad \hat{\alpha}_k^{+}=\left(\frac{\mu\omega_k}{2\hbar}\right)^{1/2}\left(\hat{Q}_{-k}-\frac{i}{\mu\omega_k}\hat{P}_k\right) \tag{23}$$

The k and $-k$ in the subscripts mean there may be both forward and backward waves simultaneously.

Since the square $|\psi(x,t)|^2$ is a real number, the operators $\hat{Q}_{-k}$ and $\hat{Q}_k^{+}$, $\hat{P}_{-k}$ and $\hat{P}_k^{+}$ obey relations

$$\hat{Q}_{-k}=\hat{Q}_k^{+},\quad \hat{P}_{-k}=\hat{P}_k^{+} \tag{24}$$

$\hat{P}_k$ and $\hat{Q}_k$ as the solutions of Equation (23) take the forms

$$P_{-k}^{+}=\hat{P}_k=i\left(\frac{\hbar\mu\omega_k}{2}\right)^{1/2}\left(\hat{\alpha}_k^{+}-\hat{\alpha}_{-k}\right),\quad \hat{P}_k^{+}=\hat{P}_{-k}=i\left(\frac{\hbar\mu\omega_k}{2}\right)^{1/2}\left(\hat{\alpha}_{-k}^{+}-\hat{\alpha}_k\right) \tag{25.1}$$

$$\hat{Q}_k^{+}=\hat{Q}_{-k}=\left(\frac{\hbar}{2\mu\omega_k}\right)^{1/2}\left(\hat{\alpha}_k^{+}+\hat{\alpha}_{-k}\right),\quad \hat{Q}_{-k}^{+}=\hat{Q}_k=\left(\frac{\hbar}{2\mu\omega_k}\right)^{1/2}\left(\hat{\alpha}_{-k}^{+}+\hat{\alpha}_k\right) \tag{25.2}$$

The operators $\hat{\alpha}_k$ and $\hat{\alpha}_k^{+}$ satisfy a commutation relation

$$\left[\hat{\alpha}_k,\hat{\alpha}_{k'}^{+}\right]=\delta_{k,k'} \tag{26}$$

The Hamiltonian is

$$\hat{H} = \sum_k \hbar\omega_k \left(\hat{\alpha}_k^+ \hat{\alpha}_k + \frac{1}{2} \right) \tag{27}$$

Comparing with the harmonic oscillator's Hamiltonian in quantum mechanics, we see that the wavefunction vibration amounts to the independent harmonic oscillators with the total number N, each of which corresponds to a mode k. In the particlenumber representation the oscillator eigenvalue equations are

$$\hat{\alpha}_k^+ \left| n_k \right\rangle = \left(n_k + 1 \right)^{1/2} \left| n_k + 1 \right\rangle, \quad \hat{\alpha}_k \left| n_k \right\rangle = \left(n_k \right)^{1/2} \left| n_k - 1 \right\rangle \tag{28}$$

where n_k is the quantum number with mode k. Equation (28) states that the $\hat{\alpha}_k^+$ is a creation operator, the $\hat{\alpha}_k$ an annihilation one; and the eigenvalue equation is

$$\hat{\alpha}_k^+ \hat{\alpha}_k \left| n_k \right\rangle = n_k \left| n_k \right\rangle \tag{29}$$

Therefore, the total energy is given by

$$E = \sum_k \hbar\omega_k \left(n_k + 1/2 \right) \tag{30}$$

Equations (29) and (30) claim that an individual wave motion with frequency ω_k, as a whole, is equivalent to boson particles of n_k, each of which has energy $\hbar\omega_k$, where $\hbar\omega_k/2$ is zero-point energy.

The μ and $\hbar$ are constant. Using Equations (18), (22), (25.2), we get a correspondence relation

$$\left|\psi_k(x,t)\right|^2 \sim \left(\frac{1}{N}\hat{\alpha}_k\hat{\alpha}_k^+\right)\Big/\omega_k \tag{31.1}$$

For a given state $\psi_k(x,t)$ with definite position x and wave vector k the ω_k in this equation is a constant determined by Equation (16). Thus, Equation (31.1) is a link between the square of the amplitude of the wavefunction $\psi_k(x,t)$ and the ratio $\hat{\alpha}_k\hat{\alpha}_k^+/N$. The eigenvalue of the operator $\hat{\alpha}_k\hat{\alpha}_k^+$ on the right side of Equation (31.1) is $n_k(x)$, which is the number of the boson particles with wave vector k at the position x due to the locality of the square of the wavefunction amplitude. The $n_k(x)$ differs from the n_k which is the total number of particles with the wave vector k in the space; the former is local, the latter is not so. Equation

(31.1) is an one-to-one correspondence relation between $n_k(x)/N$ and $\left|\psi_k(x,t)\right|^2$, the $n_k(x)/N$ is a ratio of the number of boson particles with a wave vector k at the position x against the total number of boson particles. The meanings of the number N should be emphasized here. First, there are N different positions in the region L to be provided for the electron to occupy. Second, the total number of the individual modes (or wave vectors) is N, equaling to the total number of the boson particles, since one specific mode (or wave vector) relates to one of the particles. The meaning of Equation (31.1) is explained as the following. For a uniform distribution the associated relative probability of finding the electron at each position equals to $1/N$, which amounts to that one boson particle resides in one position because there are N boson particles in total. In a nonuniform distribution, there are $n_k(x)$ times opportunities for the electron to occupy a location x, the correspondingly relative probability of finding the electron with the wave vector k at the location x is $n_k(x)/N$ tantamount to that there are $n_k(x)$ boson particles with wave vector k to lie in the same site x. From Equations (18), (22), (25.2), and (31.1), we have a corresponding relation further

$$\left|\psi(x,t)\right|^2 \sim \sum_k \left|\psi_k(x,t)\right|^2 \sim \frac{N(x)}{N} \tag{31.2}$$

where $N(x)=\sum_k n_k(x)$ is the total number of boson particles in the position x including the magnitudes of all possible wave vectors. As with Equation (31.1), the ratio $n(x)/N$ on the right side of Equation (31.2) represents the relative probability of finding the electron with all kinds of k at the position x. The particle's numbers N, $N(x)$, n_k, and $n_k(x)$ obey the following formula:

$$N = \sum_k n_k = \sum_k \int_L \mathrm{d}n_k(x) = \int_L \mathrm{d}N(x) \tag{32}$$

The normalization condition is

$$\int_L \left|\psi(x,t)\right|^2 \mathrm{d}x = \frac{1}{N}\int_L \mathrm{d}N(x) = 1 \tag{33}$$

The integral is taken in the region $L=(N-1)\cdot\varepsilon$, and the number ε tends to infinitesimal such that the distribution of the wavefunction is continuous. Equation (31) is a theoretical evidence of the probabilistic interpretation suggested by Born. The probabilistic interpretation transfers the nonlocality of an electron to the locality of the square of the wavefunction amplitude, and the wave quantization makes the locality of the square be equivalent to the locality of the boson particles.

4. DIFFERENT MEANINGS OF WAVE-PARTICLE DUALITY FOR PHOTONS AND FOR FREE ELECTRON

Double-slit interference experiment shows that the particle's behaves and the wave characteristics occur simultaneously [14] [20] . By equation (12), we get the wavefunction in the three dimensions:

$$\psi_k(\mathbf{r},t)=\psi_k(\mathbf{r})\exp(-i\omega_k t)$$

where $\psi_k(\mathbf{r})=u\exp ik\mathbf{r}$, satisfying Holmholtz equation:

$$(\nabla^2+k^2)\psi_k(\mathbf{r})=0 \tag{34}$$

Using Equation (34) and Kirchhoff's law, with the same reason as the radiation field [28] , we can confirm that the wave, like the light wave, will demonstrate single-slit diffraction, because of which the double-slit interference seems to proceed through the space containing free electron. A homogeneous wave equation is set up using the wavefunction:

$$\nabla^2\psi-\frac{k^2}{\omega^2}\frac{\partial^2\psi}{\partial t^2}=0 \tag{35}$$

If we put $k = \omega/c$, where c is the light velocity, this equation becomes

$$\nabla^2\psi - \frac{1}{c^2}\frac{\partial^2\psi}{\partial t^2} = 0 \tag{36}$$

This is just the light's wave equation in classical theory [28] . Equations (12), (34), (35), and (36) tell us that the radiation field and the space containing a free electron show some similar wave properties. Note that the wavefunctions are obtained by the classical dynamical equation, and the Equation (34) and (35) are also classical, which explicitly shows that the wave that refers to the space involving a free electron does be a classical phenomenon. For light, its wavefunction is the vector potential of the electromagnetic field, which can bring us vision effect, different from the wavefunction of the space containing a free electron. Therefore, a photon is not equivalent to a particle coming from the wave quantization of the space involving the electron, although both are bosons. In addition, double-slit interference experiment shows that these boson particles associated with the electron do not interact with the detector, except the electron. For light, the wave-particle duality refers to the same object: the light exhibits wave characteristics and the wave quantization leads to photons. For the free electron, however, the wave belong to the space containing the electron, but the electron alone, and the wave quantization leads to the boson particles distinguishable from the electron, the former are immaterial and the latter is material. The same probabilistic interpretation has different meanings for the light and for the electron. In the discussion of the probability of finding photons at a position, the wave quantization theory tells us that the number of photons at the position is usually greater than one photon, since there are N photon s in the space where the light propagates, and the number N tends to infinity. When the probabilistic interpretation describes the probability of finding a free electron at a position, the number of particles at the position is not the electron's number but the number of boson particles, which only represent the potential opportunities of the electron to occupy the position, since there is a single electron. That the huge number of the boson particles can be described by statistical theory is the reason why the probabilistic interpretation is applicable to a single electron.

5. CONCLUDING REMARK

Uncertainty relations are the intrinsic attributes of the real spaces. The bosons turning out from the wave quantizations are local in the position space because of the locality of the square of the wavefunction amplitude. These particles are immaterial and called as pseudo particles, which means that the pseudo particles may be of locality. In this sense, the Einstein's memorable remark "God does not play dice" is correct. In contrast, electron as a material particle is of nonlocality.

REFERENCES

1. Whitaker, A. (1996) Einstein, Bohr and the Quantum Dilemma. Cambridge University, Cambridge.
2. Bohr, N. (1935) Can Quantum-Mechanics Description of Physical Reality Be Considered Complete? Physical Review, 48, 696-702.
3. Hey, T. and Walters, P. (2003) The New Quantum Universe. Cambridge University, Cambridge, 165.
4. Plotnitsky, A. (2010) Epistemology and Probability: Bohr, Heisenberg, Schrödinger, and the Nature of Quantum Theoretical Thinking. Springer, New York, 114.
5. Bell, J. S. (1964) On the Einstein Podolsky Rosen Paradox. Physics, 1, 195.
6. Aspec, A., Granger, P. and Roger, G. (1981) Experimental Tests of Realistic Local Theories via Bell's Theorem. Physical Review Letters, 47, 460-463.
7. Aspect, A. (1999) Bell's Inequality Test: More Ideal than Ever. Nature, 398, 189-190.
8. Torgerson, J.R., Branning, N., Monken, C.H. and Mondel, L. (1995) Experimental Demonstration of the Violation of Local Realism without Bell Inequalities. Physics Letters A, 204, 323-328.
9. Pan J-W., Bouwmeester, D., Daniell, M., Weinfurter, H., and Zeillinger, A. (2000) Experi-mental Test of Quantum Nonlocality in Three-photon Greenberger-Horne-Zeilinger Entan-glement. Nature, 403, 515-519.

10. Born, M. (1954) The Statistical Interpretation of Quantum Mechanics. Nobel Lecture.

11. Everett, H. (1957) "Relative State" Formation of Quantum Mechanics. Reviews of Modern Physics, 29, 454.

12. Bohm, D. and Bub, J. (1966) A Proposed Solution of the Measurement Problem in Quantum Mechanics by a Hidden Variable Theory. Reviews of Modern Physics, 38, 453.

13. Omnès, R. (1992) Consistent Interpretations of Quantum Mechanics. Reviews of Modern Physics, 64, 339.

14. Zurek, W.H. (2003) Decoherence, Einselection, and the Quantum Origins of the Classical. Reviews of Modern Physics, 75, 715.

15. Einstein, A., Podolsky, B. and Rosen, N. (1935) Can Quantum-Mechanical Description of Physical Reality Be Considered Complete? Physical Review, 47, 777.

16. Schrödinger, E. (1983) The Present Situation in Quantum Mechanics. In: Wheeler, J.A. and Zurek, W.H., Eds., Quantum Theory and Measurement, Princeton University, New Jersey, 152.

17. Dirac, P.A. (1958) The Principle of Quantum Mechanics. 4th Edition, Clarendon, Oxford, 227-235.

18. Kittel, C. (1996) Introduction to Solid State Physics. 7th Edition, John Wiley & Sons, New York, 97, 450.

19. Schrödinger, E. (1933) The Fundamental Idea of Wave Mechanics. Nobel Lecture, 12 December 1933.

20. Buks, E., Schuster, R., Heiblum, M., Mahalu, D. and Umansky, V. (1998) Dephasing in Electron Interference by A "Which-Path" Detector. Nature, 391, 871-874.

21. Chen, S.S., et al. (2000) Lecture on Differential Geometry. World Scientific, Beijing, 14, 227.

22. Einstein, A. (1970) The Meaning of Relativity. 5th Edition, Princeton University, New Jersey, 35.

23. Feynman, R.P., Leighton, R.B. and Sands, M.L. (2004) Lecture on Physics, Vol. 1. Pearson Education Asia Limited and Beijing World, Beijing.

24. Feynman, R.P. and Hibbs, A.R. (1965) Quantum Mechanics and Path Integrals. McGraw-Hill, New York, 31-32, 54.

25. Amstrong, M.A. (1983) Basic Topology. Springer, New York, 39.

26. Kubo, R. (1966) The Fluctuation-Dissipation Theorem. Reports on Progress in Physics, 29, 255-284.

27. Müller-Kirsten, H.W. (2006) Introduction to Quantum Mechanics: Schrödinger Equation and Path Integral. World Scientific, New Jersey, 84.

28. Born, M. and Wolf, E. (1980) Principles of Optics. 6th Edition, Pergamon, Beijing, 81, 375-386.

INDEX